D1297731

Recursive
Programming
Techniques

Recursive Programming Techniques

WILLIAM H. BURGE
IBM Corporation
Thomas J. Watson Research Center

 ADDISON-WESLEY PUBLISHING COMPANY
Reading, Massachusetts • Menlo Park, California
London • Amsterdam • Don Mills, Ontario • Sydney

ISBN 0-201-14450-6
ABCDEFGHIJ-HA-79876

To Marianne

THE SYSTEMS PROGRAMMING SERIES

*The Program Development Process Part I—The Individual Programmer	Joel D. Aron
The Program Development Process Part II—The Programming Team	Joel D. Aron
*The Design and Structure of Programming Languages	John E. Nicholls
Mathematical Background of Programming	Frank Beckman
Structured Programming	Harlan D. Mills Richard C. Linger
The Environment for Systems Programs	Frederic G. Withington George Gardner

*An Introduction to Database Systems	C. J. Date
Compiler Engineering	Patricia Goldberg
Interactive Computer Graphics	Andries Van Dam
*Sorting and Sort Systems	Harold Lorin
Compiler Design Theory	Philip M. Lewis Daniel J. Rosenkrantz Richard E. Stearns

*Recursive Programming Techniques	William Burge
Compilers and Programming Languages	J. T. Schwartz John Cocke

*Published

Foreword

The field of systems programming primarily grew out of the efforts of many programmers and managers whose creative energy went into producing practical, utilitarian systems programs needed by the rapidly growing computer industry. Programming was practiced as an art where each programmer invented his own solutions to problems with little guidance beyond that provided by his immediate associates. In 1968, the late Ascher Opler, then at IBM, recognized that it was necessary to bring programming knowledge together in a form that would be accessible to all systems programmers. Surveying the state of the art, he decided that enough useful material existed to justify a significant codification effort. On his recommendation, IBM decided to sponsor The Systems Programming Series as a long term project to collect, organize, and publish those principles and techniques that would have lasting value throughout the industry.

The Series consists of an open-ended collection of text-reference books. The contents of each book represent the individual author's view of the subject area and do not necessarily reflect the views of the IBM Corporation. Each is organized for course use but is detailed enough for reference. Further, the Series is organized in three levels: broad introductory material in the foundation volumes, more specialized material in the software volumes, and very specialized theory in the computer science volumes. As such, the Series meets the needs of the novice, the experienced programmer, and the computer scientist.

Taken together, the Series is a record of the state of the art in systems programming that can form the technological base for the systems programming discipline.

The Editorial Board

Preface

This book describes a particular method of programming which uses a programming language based on the notation of the lambda calculus. Numerous examples of applications of this method explain and illustrate lambda-calculus techniques. The main emphasis is placed on those parts of the language, namely expressions, that denote the end results sought from the computer, rather than on the instructions which the machine must follow in order to achieve the results. The main thesis of this book is that, in many cases, this emphasis on expressions as opposed to mechanisms simplifies and improves the task of programming.

In Chapter one, the programming language used throughout the book is introduced. This language is a "sugared" version of the lambda calculus notation, and, although it does not introduce new ways of creating functions, this "sugaring" does increase the palatability of the language for practical programming purposes. This chapter also contains a brief account of the traditional approach to computability in terms of lambda-definable functions. These traditional methods of combining functions are not only valuable for exploring the limits of computability or investigating the foundations of mathematics, but also often provide the most straightforward and natural way to write a program.

Chapter two describes several solutions to the problem of constructing a programming system for converting expressions into programs. Landin's SECD machine is used as a machine model to provide mechanisms for evaluating expressions and dealing with subroutine linkage and block structure. The language can be treated as a model for the precise description of other programming languages. In order to make a more complete correspondence with other languages possible, additional features are added to the machine which correspond to assignment and **go to** statements. The resulting machine model contains, in generalized form, the corresponding features of current programming languages which may be valuable in solving practical programming problems.

Chapter three introduces a systematic way to derive programs for creating and analyzing tree-like data structures from descriptions of the sets of trees. The structure of the resulting program closely matches the data structure. This chapter also includes a technique for representing data structures by functions called "streams" such that the component parts of the structure are materialized only when they are needed. The list of tree-processing functions can be systematically converted to stream-processing functions. This technique is then used to construct programs that generate streams of combinatorial configurations.

Chapter four is concerned with methods of recognizing and parsing strings of characters. Productions of a context-free language are treated as a skeletal description of the parsing program for that language; the parsing programs are produced in a systematic way by redefining the operations on sets of character strings as functions for combining parsing programs. The structure of the parsing program that results from this redefinition matches the language structure.

Chapter five contains several examples of sorting programs expressed as recursive functions. This is often the simplest way to specify a sorting program, because the sorting of the whole collection can often be expressed in terms of the same program applied to subcollections.

The book is intended as an undergraduate or graduate text to supplement computer science courses and covers much of the material required by the ACM Curriculum Committee on Computer Science for courses on Data Structures, Programming Languages, and Compiler Construction. It is also intended as an introduction to programming technique which is particularly valuable in the early stages of the creation of a program. Most of the programs given as examples have a fairly simple structure and should provide starting points to encourage programmers to embark on more complex programming tasks along the lines suggested.

Yorktown Heights, New York   W.H.B.
May, 1975

Contents

CHAPTER 3
DATA STRUCTURES

CHAPTER 4
PARSING

CHAPTER 5
SORTING

1
Basic Notions
and
Notations

Plus ça change,
plus c'est la même chose.
ALPHONSE KARR

1.1 INTRODUCTION

This chapter contains a description of the programming language which will be used throughout this book to describe programs and programming techniques. The notation comes from several sources: some parts come from traditional mathematics; some from the more recent tradition of programming languages; while other less-familiar forms are derived from a branch of mathematical logic called *Combinatory Logic*. The meaning of these less-familiar notations will be explained in this chapter.

All the linguistic devices introduced are based upon two methods of constructing expressions from smaller expressions: 1) an operator/operand construction which denotes the application of a function to its argument, and 2) an expression format which denotes a function using Church's lambda notation. Thus the extra notation that is added to this basis adds no new structural features because each new piece of notation can be rewritten in terms of these two constructions, with the possible addition of a few constants. These constants are common to most programming applications, and correspond to conditional expressions, lists of expressions, and self-referential definitions. The additions create a practical and powerful programming system, which is more like a *family* of programming languages than a single language, because the features introduced are concerned more with combining functions to produce new ones than with the nature of the primitive functions that are being combined. A programming language for a particular range of applications can be obtained by adding an appropriate set of primitives to this basic structure.

1

The programming system is first described informally, and examples of its use are given. The features introduced are next described more formally by a translator to Church's lambda notation. [1–6] In the calculus of lambda conversion, the calculation of the result of applying a function is represented by successive transformations of an expression to one having a simpler form. A brief account of this method of evaluating an expression is given together with a demonstration of how the partial recursive functions can be represented by using this notation.

Emphasis is placed on the expression parts of a programming language because they have a property that is very valuable for practical programming purposes, namely that *the value, or meaning, of an expression depends in a simple way only on the values of its subexpressions and on no other properties of them.* This property enables a programmer to master a complex program by splitting it into simpler independent programs. In other words it is possible to make the structure of the program match the structure of the problem being solved.

1.2   OPERATOR/OPERAND EXPRESSIONS

Functions and types.   A function is a rule of correspondence by which when one thing is given as an argument, another unique thing may be obtained called the *value of the function for that argument.* Another way of expressing this is to say that a function *f converts* or *transforms* x to y, or that y is the *result of the application* of f to x. The result of applying f to x will be written $f(x)$ or $f\,x$. This is a composite expression made up of an *operator* part f and an *operand* part x. In general, the standard way of indicating the result of an application is to construct an expression from two expressions by enclosing one or both within parentheses and by placing the operator part in front of the operand part. The resulting expression will be said to be *composed by application.*

Many functions that will be introduced may only be applied to certain kinds of objects and will only produce objects of a certain type. The set of objects to which a function may be applied will be called its *domain,* and the set of objects that can result from applying the function will be called its *range.* The type of a function will be written as follows:

$$A \to B$$

which indicates that when the function is applied to a member of the set A it produces a member of the set B as a result. The fact that a function for squaring an integer produces an integer, for instance, can be expressed

$$square \; \varepsilon \; (\text{integer} \to \text{integer}).$$

Other examples are

$$\sin \varepsilon \ (\text{real} \rightarrow \text{real})$$
$$\log_e \varepsilon \ (\text{positive} \rightarrow \text{real})$$
$$\text{negating } \varepsilon \ (\text{positive} \rightarrow \text{negative}).$$

In general an assertion of the form

$$g \ \varepsilon \ (A \rightarrow B)$$

implies that A and B are two sets and that g denotes a function whose domain includes A and associates a member of B with every member of A. It is important to distinguish between f, for example, as an identifier that denotes a function, and $f(x)$ as an expression that denotes the result of applying f to x. The three expressions f, x, and $f(x)$ are related as follows. If A and B are two sets, and if $x \ \varepsilon \ A$ and $f \varepsilon \ (A \rightarrow B)$ then it can be deduced that

$$f(x) \ \varepsilon \ B.$$

A function of two arguments is characterized as being applicable to two arguments in a certain order and yielding, when so applied, a certain result. If the function is applicable to an ordered pair of arguments whose first item is a member of the set A and whose second item is a member of the set B and if it produces as result a member of the set C, then the type of the function can be indicated by $A \times B \rightarrow C$. Two examples follow:

$$+ \ \varepsilon \ (\text{real} \times \text{real} \rightarrow \text{real})$$
$$> \ \varepsilon \ (\text{real} \times \text{real} \rightarrow \text{truth value})$$

The result of applying a function f to two arguments will be written $f(x, y)$. An ordered pair of arguments will be treated as a special case of a list of arguments. In general, if $A_1, A_2, \dots, A_n$ are sets, then the expression:

$$A_1 \times A_2 \times \dots \times A_n,$$

known as the *Cartesian product* of the sets, will be used to denote the set of all n-tuples whose first, second, third, etc. items are members of A_1, A_2, A_3, etc., respectively.

A function whose domain and range are the same set will sometimes be called a *transformer,* and a function whose result is a *truth value* will be called a *predicate.* It follows that an *A-transformer* is a function of the type $(A \rightarrow A)$, and an *A-predicate* has the type $(A \rightarrow \text{truthvalue})$.

There are many other ways of indicating the applicative structure of an expression. The operator/operand structure has to be uncovered in order to find the value of an expression. For instance, the expression $3x^2 + y$ can be characterized as having $+$ as its operator, and $3x^2$ and y as its two operands. The subexpression $3x^2$ has $\times$ as its operator; and its operands are 3 and the expression, x^2, whose operator is the squaring function and whose operand is x. The following rules had to be used to determine the applicative structure:

- The *infixed operator,* $+$, is placed between its two operands.
- In $3x^2$ the multiplication operator is implicit.
- The $+$ rather than the multiplication or the squaring is the operator.
- The ^{2} is a *postfixed operator* for squaring.

The operator/operand structure can be exhibited more clearly by rewriting the expression in prefix notation as follows:

$$+ \ (\ \times \ (3, \ square(x)), \ y)$$

In prefix notation the problems of disentangling the operator and operand parts of an expression have been simplified because the structure is more clearly apparent. In order to be able to construe any piece of notation as an expression, the following questions must be answered for both the operator and operand expressions and for the complete expression.

1. *Is the expression simple or composite?*
2. *If it is composite, what is its operator part, and what is its operand part?*

If these questions can be answered then it may be possible to find the value of the expression. The smallest subexpressions in $3x^2 + y$ are $x, y, +, \times,$ and *square,* and are all examples of *identifiers.* An identifier will be written as any number of letters or digits, or as a special symbol such as $+$. The fact that multicharacter identifiers are going to be used means that two adjacent identifiers must be separated by a space.

Value and denotation.   The value of an expression depends *only* on the values of its subexpressions. The value of an identifier is the value that has been assigned to it, and an identifier with a known value, such as $+$, will be called a *constant.* The value of a composite expression is found by applying the value of the operator part of the expression (which must be a function) to the value of the operand. The words operator and operand are reserved to indicate expressions rather than their values. The value of an expression might be a number, truth value, data structure, function, or any other object that can be represented inside a computer. A function bears the same relation to an operator as a number does to a numerical expression. The function itself consists of the determination, or yielding, of a value from each argument in the domain of the function. The expression will be said to denote its value, and the value is called the *denotation* of the expression. Two expressions having the same value are said to be equivalent. Since the value of an expression depends only on the values of its subexpressions the replacement of any subexpression by an equivalent expression produces an equivalent expression.

Function valued functions.   A notation that has become established in combinatory logic will next be introduced. In this notation the operator may be a composite expression. Although functions that produce a function as a result of their application do occur in mathematics, there is usually a special notation reserved for each function. In $log_n x$, for example, the *log* part may be considered to denote a function which when applied to a positive number n, produces the function log_n, and has the type

$$(\text{positive} \rightarrow (\text{positive} \rightarrow \text{real}))$$

In other words, *log* operates on a positive real number to produce a function from positive real numbers to real numbers.

The differential operator is an example of a function that both operates on and produces a function. Another example of such a function is one that applies a given function twice. Associated with each transformer f there is another function that first applies f and then applies f to the result. Given the function for squaring, for instance, this rule produces the function for finding the fourth power of a number. The function *twice* can be defined as

$$twice \, f \, x = f(f \, x).$$

When *twice* operates on an $(A \rightarrow A)$ type function f and an object of type A, namely x, it produces an object of type A. Alternatively *twice* can be thought of as operating on a function of type $(A \rightarrow A)$ and producing a function of the same type. The type of *twice* is

$$(A \rightarrow A) \rightarrow (A \rightarrow A).$$

A third example of a function-producing function operates on two functions to produce a third. If the range of a function f is the same as the domain of a function g then it is possible to form the *functional product* of g and f, written $g \cdot f$, such that $(g \cdot f)x = g(f \, x)$, in which f is first applied to the argument, and g is then applied to the result. If $f \, \varepsilon \, (A \rightarrow B)$ and $g \, \varepsilon \, (B \rightarrow C)$ then $g \cdot f \, \varepsilon \, (A \rightarrow C)$. In other words, the type of the infixed point is

$$((A \rightarrow B) \times (B \rightarrow C)) \rightarrow (A \rightarrow C).$$

The functional product is associative and so expressions like $f \cdot g \cdot h$ are unambiguous. The n-fold application of a transformer f, will be written f^n, and since $f^n \cdot f^m = f^{n+m}$ it is appropriate to define $f^1 = f$, and f^0 as the identity function I, such that $I\,x = x$.

As has been seen, an expression can have a composite expression as its operator part, so that it is sensible to write expressions like $(f(x))(y)$ whose value is obtained by first applying f to x, producing a function which is then applied to y. Parentheses in expressions are used solely to group together subexpressions, and certain conventions will be introduced for omitting them. The parentheses in $f(x)$, for example, can always be omitted, and so $f\,x$ means the same thing. By convention all the parentheses in $(f(x))(y)$ may be omitted, and the expression can be rewritten $f\,x\,y$. In other words, in a sequence of subexpressions separation is by an implicit apply operator that is assumed to be bracketed to the left. So the expression $a\,b\,c\,d\,e$ will be bracketed $(((ab)c)d)e$.

This leads to another way in which a function can be said to have more than one argument. The application of a function to several arguments can be done one argument at a time, so that $f\,a\,b$ is $(f\,a)b$ and $f\,a\,b\,c$ is $((f\,a)b)c$, etc. If p denotes an addition function of this sort then the expression $a + b$ could be rendered as $p\,a\,b$. Although this may seem to be a mere change of notation for theoretical purposes, or to aid a mechanical processor, certain practical programming benefits may arise from using this alternative. In using $p\,a\,b$ rather than $a + b$ the programmer is able to denote three objects: 1) p, the addition function, 2) $(p\,a)$ the function that adds a to a number, and 3) $(p\,a\,b)$, the sum. Using the infixed plus operator he can only denote the sum.

Since the language is intended for practical use, as well as for explaining programming concepts; both ways of defining multiargument functions are retained. Essentially the same function may appear in two guises as, for example, $(f\,x\,y)$ and $f(x, y)$. These will be treated as two different functions, and both will be treated as having one argument. In the first case f is applied to x, and the resulting function is applied to y. In the second case f is applied to one argument which is an ordered pair. In other words the type of the first function is $(A \rightarrow B) \rightarrow C$, and the type of the second function is $(A \times B) \rightarrow C$. There are two ways in which the expression $a + b - c + f(a, c)$ can be put into strict operator/operand form:

1. by using ordered pair notation, producing

$$+ (-(+ (a, b), c), f(a, c));$$

2. by defining p, m, and g as

$$p\,x\,y = x + y$$
$$m\,x\,y = x - y$$
$$g\,x\,y = f(x, y)$$

and rewriting it as $p\ (m(p\,a\,b)c)(g\,a\,c)$.

Both the infixed and prefixed forms of the operator will be used in this book. We will adopt the convention that application has a stronger precedence than any infixed operator, and that a comma has a weaker precedence than any infixed operator, so that $f x + g y$, for example, will be bracketed $(f x) + (g y)$ and $a + b, c + d$ will be bracketed $(a + b), (c + d)$.

To summarize this section: The structure of the expressions has been simplified (with some loss of legibility), so that it is completely exposed if the following questions can be answered for the complete expression and for both the operator and operand portions.

1. *Is it simple or composite?*
2. *If simple, what is the identifier?*
3. *If composite, what are the operator and operand parts?*

1.3   VARIABLES AND LAMBDA EXPRESSIONS

Variables are often used in both mathematics and programming. In

$$\Sigma_{i=1}{}^{n} A_i \quad \text{and} \quad \int x^2 + x.dx,$$

the i in the first expression and the x in the second expression are examples of variables. They are irrelevant to the meaning of the whole expression and may be replaced by j and y respectively, without changing the meaning of the complete expressions, giving the equivalents

$$\Sigma_{j=1}{}^{n} A_j \quad \text{and} \quad \int y^2 + y.dy.$$

In terms closer to programming the choice of x in the following definition of a function,

$$f(x) = 10x^2 + 4x + 3$$

is also unimportant in defining f and could be replaced on both sides by y.

A definition such as this has three parts: 1) the name of the function being defined, namely f, which will be called the *definiendum, or definee;* 2) the variable x; and 3) the expression on the right hand side which determines the value of the function for each argument. This last part is sometimes called the *associated form* of the function.

In Church's lambda notation, it is possible to write an expression, called a *lambda expression,* which denotes a function. Using this notation, the expression

$$\lambda x.10x^2 + 4x + 3$$

denotes the function that is associated with f in the definition above. The same definition could equally well be written

$$f = \lambda x.10x^2 + 4x + 3.$$

A lambda expression of the form $\lambda x.M$ has two parts. The part between the lambda and the point will be called its *bound variable,* and the part following the point will be called its *body.* The lambda expression that results from prefixing the $\lambda x.$ to an associated form M will be said to be *composed by abstraction.* It is important to distinguish between a lambda expression and an associated form. Whereas the value of the expression $10x^2 + 4x + 3$ is a number which can only be obtained when the value of x is known; the value of $\lambda x.10x^2 + 4x + 3$ is a function. In other words the type of the value of $10x^2 + 4x + 3$ is real, whereas the value of $\lambda x.10x^2 + 4x + 3$ is of type (real $\rightarrow$ real).

The right hand side of a definition is not the only context in which a lambda expression can occur. An expression denoting the application of the function above to 5 could be written either $f(5)$ or $(\lambda x.10x^2 + 4x + 3)5$. The first expression relies on the definition of f given elsewhere; the second expression contains all the information needed to produce the result. A lambda expression can also occur in operand position, and

$$(\lambda f.f(5))(\lambda x.10x^2 + 4x + 3)$$

means the same as the two expressions above.

A function of two arguments can be defined as

$$g(y) = \lambda x.10x^2 + 4xy + 3y^2$$

in which the right hand side is a lambda expression. When g is applied to a number it produces a function as a result. For instance, if g is applied to 1 it produces the function f, defined above. By rewriting this definition of g using the same rule as above for moving variables from one side of the equality to the other the same function could be defined by

$$g(y)(x) = 10x^2 + 4xy + 3y^2$$

or as

$$g = \lambda y.\lambda x.10x^2 + 4xy + 3y^2.$$

This last definition which has a single identifier on the left hand side will be called a *definition in standard form.* By convention, parentheses may also be omitted from the variables on the left side of definitions. In fact, it is never necessary to enclose a single identifier in parentheses, and the following definitions are other alternative forms:

$$g\,y = \lambda x.10x^2 + 4xy + 3y^2$$
$$g\,y\,x = 10x^2 + 4xy + 3y^2$$

An expression may occur in three positions as a component of a larger expression:

1. in the operator position,
2. in the operand position,
3. as the body of another lambda expression.

The lambda expression is the second basic method of assembling a new expression. In their most austere form the expressions under consideration may be characterized as follows.

- An expression is
  either *simple* and is an identifier
  or a *lambda expression*
       and has a *bound variable* which is an identifier
       and a *body* which is an expression,
  or it is *composite*
       and has an *operator* and an *operand,* both of which are expressions.

A rule is needed for recognizing when the body of a lambda expression ends. The rule is that the body extends as far as it can until it is terminated by a closing bracket, a comma, or the end of the whole expression. It follows that parentheses are only needed to enclose the body if it is a list although they may be used if this improves readability. Parentheses that enclose the whole lambda expression are unnecessary in a context in which it is followed by a closing bracket or comma. Thus $(\lambda x.\lambda y.x + y)a$ means the same as $(\lambda x.(\lambda y.x + y))a$, and $\lambda x.x + y, x$ means the same as $(\lambda x.x + y), x$.

Another abbreviation is possible when a lambda expression occurs as the body of another lambda expression. In this case $\lambda x.\lambda y.x + y$ may be written $\lambda x \, y.x + y$ and $\lambda x.\lambda y.\lambda z.x + y + z$ may be written

$$\lambda x \, y \, z.x + y + z.$$

1.4   DATA STRUCTURES

In the following pages many different data structures are used. This section describes the method which will be employed to define new data structures. The method specifies the *abstract syntax* of sets of data structures, as opposed to the *concrete syntax* which refers to the set of strings of characters which define a written representation. Each structure description introduces functions for both constructing and accessing the component parts of members of the set described. It is a method of introducing new data types without being too precisely committed to the way they are represented inside the computer. The functions which are introduced in a data definition are limited only by certain axioms that relate them. A new structure will be defined using an English sentence of the form

- A pair has a *first* and a *second.*

If x is a pair then (*first x*) denotes the first of the pair, and (*second x*) denotes the second of the pair. The definition explicitly names the selecting functions, or *selectors,* for the parts of a structure but gives no name to the function for constructing pairs, which must be given separately. In the case of a pair it will be called *cpair.* This definition brings three functions called *first, second,* and *cpair* into existence. In order to mechanize programs which contain references to these functions they must be implemented by programs. The only properties required of these programs, however, are that:

$$first(cpair \ x \ y) \ = \ x$$
$$second(cpair \ x \ y) \ = \ y,$$

from which it follows that if z is a pair then

$$cpair(first \ z)(second \ z) \ = \ z.$$

All the functions that are introduced by definitions of data structures will be assumed to satisfy these kinds of axioms. A set of data structures might be made up of two or more sets having different formats. A list can be defined as having one of two different formats. It is either *null,* and has no components, or *nonnull* and has two components, a head and a tail which is also a list.

These definitions may be made more precise by specifying the nature of the component data structures. An *A-B-pair,* for example, can be defined to be a pair whose first is an object of type A (or which belongs to the set A) and whose second is an object of type B, as follows:

- An *A-B-pair*

    has a *first* which is an A,

    and a *second* which is a B.

It is then sensible to write *integer-list-pair* to denote the set of pairs whose first is an integer and whose second is a list. Similarly

- An *A-list* either

    is *null*

    or has a *head* which is an A and a *tail* which is an *A-list.*

defines a list whose elements all belong to the set A. It is then sensible to write *integer-list* as the name of the set of lists whose elements are integers.

When a data structure has more than one format, there are associated predicates that distinguish between different formats. In the case of a list there is one predicate which is called *null*. When *null* is applied to a null list it returns the value **true,** when applied to a nonnull list it returns the value **false.** The functions *head* and *tail* are called selectors. The functions which manufacture structures from their components are called *constructors.* The constructor for nonnull lists is called *prefix,* and the null list is written (). The types of the operations on lists are:

$$null \; \varepsilon \; A\text{-list} \rightarrow \text{truthvalue}$$
$$head \; \varepsilon \; (\text{nonnull } A\text{-list} \rightarrow A \,)$$
$$tail \; \varepsilon \; (\text{nonnull } A\text{-list} \rightarrow A\text{-list})$$
$$prefix \; \epsilon \; (A \; \rightarrow \; (A\text{-list} \rightarrow A\text{-list}))$$

These functions are closely interrelated as follows:

$$null() = \textbf{true}$$
$$null(prefix \; x \; y) = \textbf{false}$$
$$head(prefix \; x \; y) = x$$
$$tail(prefix \; x \; y) = y$$
$$prefix(head \; z)(tail \; z) = z$$

where x is an A, y is an A-list, and z is a nonnull A-list.

Lists occur so frequently in the following pages that special notations are used for them. First, *head* and *tail* will be abbreviated to h and t. Second, $x{:}y$ is used as an alternative method of writing *prefix x y*. Also when $k \geq 2$,

$$x_1, x_2, x_3, \ldots, x_k$$

is used as a syntactic variant of

$$x_1 : (x_2 : (x_3 : (\ldots x_k : () \ldots))).$$

By convention, the *head* of a list is always written at the left. A list of length k is called a k-list, and the list selectors for the first, second, third,...etc. members of the list are called *first, second, third,* etc. In general nth is written for $h.t^{n-1}$. Lists of length one, like $(x{:}())$, are written $u \; x$, in which u is a function that transforms an object into a list of length one which contains the object. Another data structure that is often used is a *list structure,* defined below, whose list elements may themselves be lists.

- An *A-list structure* either

    is *atomic* and is an A

    or else it is *not atomic* and is an (A-*list structure*)-*list.*

The same operations can be used for lists and list structures. The *atomic* predicate is the only extra operation introduced by this last definition. An example of the notation to be used for list structures is

$$(a, b), ((c, d), u\ e), ()$$

which denotes a list of length 3 whose first element is (a, b), whose second element is $((c, d), u\ e)$, and whose third item is ().
Examples of other data structures that have already been discussed are:

- A type either
    is *simple* and is an identifier,
    or has a *left* and a *right* both of which are types.
- An AE (applicative expression) is:
    simple and is an identifier,
    or is a *lambda expression*
        and has a *bound variable* which is an identifier
        and a *body* which is an AE,
    or is compound and
        has an *operator,*
        and an *operand* both of which are AE's.

1.5   ADDITIONAL EXPRESSION FORMS

List structured bound variables.   The syntax of lambda expressions can be extended by permitting identifier-list structures to occupy their bound variable positions. For example

$$\lambda(x, y, z).f(x, g(z, y))$$
$$\lambda(x, y).f(x, g(z, y))$$
$$\lambda(u\ x).f(x, g(z, y))$$
$$\lambda().f(x, g(z, y))$$

denote functions whose domains are limited to lists of length 3, 2, 1, and 0, respectively. Within their domains they are equivalent to

$$\lambda s.f(first\ s, g(third\ s, second\ s))$$
$$\lambda s.f(first\ s, g(z, second\ s))$$
$$\lambda s.f(first\ s, g(z, y))$$
$$\lambda s.f(x, g(z, y)).$$

Again, the expression

$$\lambda(w, (x, y), z).(f(w, y), g(x, z))$$

denotes a function which is applicable to a 3-list whose second item is a 2-list. Within this domain it is equivalent to

$$\lambda s.(f(first\ s,\ second(second\ s)),\ g(first\ (second\ s),\ third\ s)).$$

Definitions of functions may also have list structures of identifiers on the left hand side. For example

$$f(x,\ y,\ z) = x^2 + y^2 + z^2$$

which means the same as

$$f = \lambda(x,\ y,\ z).x^2 + y^2 + z^2.$$

A list of expressions is also an expression, and is called a *listing*. No meaning is given to a list in operator position. The value of a list of expressions is a list of the same size in which each expression has been replaced by its value. It is possible to define functions that produce a list as a result. For example

$$cxplus(x,\ y)(s,\ t) = x + s,\ y + t$$

in which the result of applying the function is a pair, or list of length two. The type of the function *cxplus* is

$$(real \times real) \rightarrow ((real \times real) \rightarrow (real \times real))$$

Although a function, like *f* above, may be defined as expecting an argument that is a 3-list; it need not necessarily be applied to an operand that is a listing, like expression of the style $f(3, 5, 2)$. It is only necessary that the operand be an expression whose value is a 3-list. The expressions (*first* ((3, 5, 2), 6), and (3:(5:(2:()))), for example, are other possible forms that the argument of *f* could take.

The syntax of the expressions which have been introduced so far can be characterized as follows:

■   An expression is
      an *identifier*
      or a *lambda expression*
         and has a *bound variable* (*bv*)
            which is an *identifier*-list structure
         and a *body* which is an expression,
      or is a *listing* and is an expression-list,
      or is *compound*
         and has an *operator*
         and an *operand,* both of which are expressions.

These list-structured bound variables serve two purposes: 1) they impose a constraint upon the domain of applicability of the function; and 2) they name the components of the list structure, rather than the complete structure. The syntax of the bound variable can be elaborated in other ways. It is possible to denote functions that are restricted to operate on other data structures. The expression

$$\lambda(x:y).x:x:y$$

denotes a function whose domain is limited to nonnull lists. The *head* of the list is named x and its *tail* is named y. The function prefixes a copy of the *head* of the list to the original list. In general the bound variable part has to be interpreted as a data structure containing distinct identifiers and it must be possible to derive the chain of selectors leading to each identifier. Another such elaboration is similar to a programming language declaration. A type is attached to a variable and constrains the function to be applicable only to that type of object. The expression

$$\lambda \; integer \; x.x^2 + 4x + 3,$$

for example, denotes a function whose domain is restricted to integers.

Auxiliary definitions   The practice of first naming a value and then later using it by referring to its name is such a valuable way of decomposing a calculation that it warrants a special syntax. The corresponding construction in the lambda notation is a compound expression whose operator is a lambda expression. Other methods of writing this construction are either

$$M \quad \textbf{where} \quad x = N$$

or

$$\textbf{let} \; x = N \; M$$

instead of

$$(\lambda x.M)N$$

because the alternatives are frequently easier to read and understand. These formats are most useful when the auxiliary definition is used more than once, for example

$$(x + 3)(x - 2) \quad \textbf{where} \quad x = ay^2 + by + c.$$

The *where* construction corresponds to "top down programming" in which objects are first assumed to exist and later defined. The *let* construction corresponds to "bottom up programming," in which objects are defined first and used later. Sometimes a semicolon will be used to separate N and M

in the *let* expression, and sometimes indentation will be used to make the structure of the written expression clear. So

$$\textbf{let } x = M; N$$

$$\textbf{let } x = M$$
$$N$$

$$M$$
$$\textbf{where} \quad x = N$$

are alternative ways to qualify an expression with an auxiliary definition. The only indentation rule used to resolve ambiguities is that the whole subexpression lies in the southeast quadrant of the page defined by its first character.

Qualified expressions have two parts:

1. the expression M, called the main expression,
2. an auxiliary definition taking either of the forms

$$\textbf{where } x = N \qquad \text{or} \qquad \textbf{let } x = N.$$

The following case must be added to the structure descriptions for expressions.

■   or is *qualified* and

has a *main* which is an expression

and an *auxiliary* which is a definition,

where a definition has a *definee* which is an bound variable

and a *definiens* which is an expression.

The definition above is in standard form but definitions can have other formats. It is possible to qualify expressions by definitions of functions, for example,

$$f(3) + f(f(7)) \qquad \textbf{where} \qquad f(x) = x^2 + 3.$$

In general,

$$M \qquad \textbf{where} \qquad f(x) = N$$

can be reduced to operator/operand form in two stages; first to:

$$M \qquad \textbf{where} \qquad f = \lambda x.N$$

then to

$$(\lambda f.M)(\lambda x.N).$$

Another case has to be added to the definition format, producing the structural description:

- A definition either is *standard*

    and has a *definee* which is an bound variable

    and a *definiens* which is an expression

  or is *functional*

    and has a *leftside* which is a nonnull bound-variable-list

    and a *rightside* which is an expression.

The functional definition is transformed to a standard definition according to the following example. First

$$f\,x\,y\,z = N$$

is transformed to

$$f = \lambda x\,y\,z.N$$

and then to

$$f = \lambda x.\lambda y.\lambda z.N$$

A similar device will be used for lambda expressions having list structured bound variables such as

$$(\lambda(x, y, z).M)(R, S, T)$$

which may be written in any of the three following ways:

1. *M* **where** $x, y, z = R, S, T$

2. *M* **where** $x = R$
    and $y = S$
    and $z = T$

3. **let** $x = R$
    and $y = S$
    and $z = T$
        M.

The last two formats permit an expression to be qualified by more than one definition simultaneously. The additional definition format may be written

  or is *simultaneous* and is a definition-list.

Examples of simultaneous definitions are:

$$ax^2 + bx + c \quad \textbf{where} \quad (a, b, c, x) = (4, 3, 6, 1)$$
$$\textbf{let } f(x) = x^2 + 5$$
$$\textbf{and } y = 77$$
$$f(y^2) - 24y.$$

In general, the expression that results from a simultaneous definition is formed by first translating each auxiliary definition to standard form, then assembling the definees into a list to form the definee, and finally assembling the definiens into a list to form the new definiens.

Conditional expressions. A conditional expression is written

$$\textbf{if } P \textbf{ then } A \textbf{ else } B$$

or it can be written in indented forms such as

$$\textbf{if } P$$
$$\textbf{then } A$$
$$\textbf{else } B$$

$$\textbf{if } P \textbf{ then } A$$
$$\textbf{else } B.$$

This expression has three component expressions, P, A, and B. P is an expression that denotes a truth value. When its value is **true,** the value of the whole expression is the value of A, otherwise it is the value of B. It is important to evaluate the expression P before choosing which of A or B has to be evaluated. This means that the expression A only needs to be given a meaning when P is **true,** and B only needs to be given a meaning when the value of P is **false.** This enables conditional expressions like

$$\textbf{if } x = 0 \textbf{ then } 1 \textbf{ else } 1/x$$

to be written in which $1/x$ is only to be evaluated when x is nonzero.

The conditional expression cannot be mirrored by the application of a function to the list (P, A, B) because the function is applied to the values of the expressions, P, A, and B. It can, however, be reduced to operator/operand form at the expense of introducing a function called *if* which operates on a truth value and produces the selector first if **true** and second if **false.** The conditional expression may be translated to

$$((\textit{if } P)(\lambda().A, \lambda().B))().$$

The function $\lambda().A$ is only applicable to the null list, and when it is applied it produces the value of A. This is because the body of a lambda expression is only evaluated when the lambda expression is applied. When A is replaced by $\lambda().A$ it has the effect of delaying the evaluation of A until the function is applied. The expression $(if\ P)$ produces a selector which selects the function to be applied to the null list. It follows that only one of A or B is evaluated. A conditional expression having the form

if P

then A

will have an undefined value if the condition is not satisfied.

The extra formats to be added for conditional expressions are as follows.

■   or is a *conditional* and has
>   a *condition* which is an expression
>   and is either *two-armed*, with
>>   a *leftarm* and a *rightarm*
>>>   both of which are expressions,
>>   or is *one-armed*
>>   and has an *arm* which is an expression.

A similar construction is a *case expression* in which a list of expressions is used instead of a pair. This construction may be written

case E of $(A, B, C, D, \ldots)$

and means that the evaluations of $A, B, C, \ldots$, etc. are to be delayed until E has been evaluated to produce a list selector, *first,second,third* etc. The case expression may be rewritten

$E(\lambda().A, \lambda().C, \ldots)()$.

Circular definitions.   It is difficult at first sight to see how programs with loops can be written using expressions. A program with a loop, however, corresponds to a *circular* or *self-referential* definition of a function. A circular definition of a function defines the result of applying the function in terms of the result of applying the same function to a simpler argument. In the terminology of computer science these are known as *recursive definitions,* although in recursive functon theory the adjective *recursive* refers both to circular definitions and to definitions of functions in terms of functions which are in some sense simpler. An example of a circular definition is:

$$s(n,k) =$$
$$\text{if } k = 1$$
$$\text{then } 1$$
$$\text{else if } k = n$$
$$\quad\text{then } I$$
$$\quad\text{else } s(n-1, k-1) + k \times s(n-1, k)$$

which is the definition of Stirling numbers of the second kind. The evaluation of $s(4, 2)$, for example, proceeds as follows.

$$s(4, 2) = s(3, 1) + 2 \times s(3, 2)$$
$$= 1 + 2 \times (s(2, 1) + 2 \times s(2, 2))$$
$$= 1 + 2 \times (1 + 2 \times 1)$$
$$= 7$$

Other examples of circular definitions are:

$$length\ x =$$
$$\text{if } null\ x$$
$$\text{then } 0$$
$$\text{else } 1 + length(t\ x)$$
$$factorial\ x =$$
$$\text{if } zero\ x$$
$$\text{then } 1$$
$$\text{else } n \times factorial(n - 1)$$

It is possible to transform a circular definition into a standard definition, i.e., one whose definee is an identifier and whose definiens is an expression. This can be done by constructing an expression that denotes a selfreferential function by using a *fixed-point-finding* function called Y. The characteristic property of Y, *namely that* $Y f = f(Y f)$, will be assumed for the moment. Several possible implementations of Y will be given later.

A self-referential definition of a function f can be rearranged as a standard definition of the form $f = E f$ in which the expression E does not contain f. One solution of this equation is $(Y E)$. The definition of *length*, given above, can be rewritten in stages as follows:

$$length = \lambda x. \text{ if } null \ x$$
$$\textbf{then } 0$$
$$\textbf{else } 1 + length(t \ x)$$

The identifier *length* can next be abstracted from the right hand side to give:

$$length = (\lambda f.\lambda x.\text{if } null \ x$$
$$\textbf{then } 0$$
$$\textbf{else } 1 + f(t \ x)) \ length$$

This may now be rewritten using *Y*.

$$length = Y(\lambda f.\lambda x.\text{if } null \ x$$
$$\textbf{then } 0$$
$$\textbf{else } 1 + f(t \ x))$$

The right-hand side of this standard-form definition is an expression for the function *length* which contains no occurrences of the identifier *length*. The definition of a circular function will be written with the piece of syntax **rec** preceding the definition. The new format to be added to definitions is

or is a *circular definition* and has a *body* which is a definition.

The equivalent definition is found by reducing the body of a circular definition to standard form, $x = M$ and then by forming the new standard definition $x = Y\lambda x.M$.

The function *Y* may also be applied to list transforming functions. The circular list *L*, defined

$$L = (1:2:3:L),$$

associates with *L* the list whose first is 1, second is 2, third is 3, fourth is 1, etc. It is a device for defining the infinite periodic list

$$L = 1, 2, 3, 1, 2, 3, 1, 2, 3, 1, 2, 3, \ldots$$

and the natural way to represent such a list is by the looped structure in Fig. 1.1.

The list *L* is denoted by the expression

$$Y\lambda L.(1:2:3:L).$$

When *Y* is applied to a function that transforms one list of functions to another list of functions, the results are known as *mutually recursive* functions. If *f*, *g*, and *h* are three mutually recursive functions then any occur-

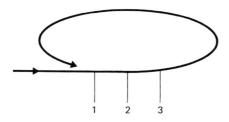

Fig. 1.1   A looped list.

rence of the identifiers f, g, and h in the definiens of these functions refer to the functions being defined, rather than to any other meanings that might be associated with them.

An example of the circular definition of three mutually recursive functions follows.

$$\textbf{rec } f(x) \; = \; \ldots g \ldots h \ldots f \ldots g$$
$$\textbf{and } g(y) \; = \; \ldots h \ldots f \ldots f \ldots h$$
$$\textbf{and } h(z) \; = \; \ldots f \ldots g \ldots h \ldots g$$

In these a simultaneous definition is preceded by **rec**. The definitions are first translated to standard form, i.e.,

$$\textbf{rec } (f, g, h) = ((\lambda x \ldots g \ldots h \ldots f \ldots g), (\lambda y \ldots h \ldots f \ldots f \ldots h),$$
$$(\lambda z \ldots f \ldots g \ldots h \ldots g))$$

and then to

$$(f, g, h) = Y\lambda (f, g, h).((\lambda x \ldots g \ldots h \ldots f \ldots g), (\lambda y \ldots h \ldots f \ldots h),$$
$$(\lambda z \ldots f \ldots g \ldots h \ldots g)).$$

The bound variable list can be translated to a single identifier as before.

1.6   EXAMPLES

This section includes a number of examples showing how programs can be written using the notation introduced above. These examples are written in the form of a series of messages which are supposed to be addressed to a programming system. Two types of messages are allowed. The first is a definition message, and will be preceded by the word **def.** The system makes no printed response to a definition, but stores it for future use by adding it to a list of name-value pairs. The second type of message is an expression, and the system responds by printing its value. At the start, the system is assumed to know the meaning of the arithmetical operators $+$, $-$, $\times$; the relational operators $<$, $=$, and $>$; the functional composition operator $\cdot$; the list operators *prefix, null, h,* and *t:* the empty list (): and the list structure predicate, *atomic.*

$2 + 2$
4

def $x = 2$

$x + x$
4

def $f\,x = x + x$
$f\,2$
4

$f(f\,2)$
8

let $g\,x = x + x\,;g\,2$
4

def *plus* $x\,y = x + y$
def *mult* $x\,y = x \times y$
def *minus* $x\,y = y - x$

Prefix notation

plus 2 2
4

(*plus* 2)2
4

(*plus*(2))(2)
4

def $g\,y = y \times y$
$g\,2$
4

def $W\,f\,x = f\,x\,x$
def *double* $= W\,plus$

The definiens can denote a function.

double 2
4

def *square* $= W\,mult$

square 2
4

def *twice* $f = f\cdot f$
Functional composition.

twice square 2
16

def *thrice* $f = f\cdot f\cdot f$
thrice square 2
256

def $v(x, y) =$
$\quad x^2 + y^2, x^2 - y^2$

$v(3, 2)$
13, 5

Both the argument and result can be a list.

$v(v(3, 2))$
194,144

def *greater* $x\,y = y > x$

(*greater* x) is **true** when applied to a number greater than x.

def *less* $x\,y = y < x$
def *max* $x\,y =$
$\quad$ **if** $x < y$
$\quad$ **then** x
$\quad$ **else** y
greater 2 3
true

max 4 6
6

def *div* $m\,n =$
$\quad$ **if** $m < n$
$\quad$ **then** 0, m
$\quad$ **else**
$\quad$ **let** $x, y = div(m - n)n$
$\qquad\qquad x + 1, y$

div 7 2
3,1
  def *divides x y* =
      (*second*(*div x y*) = 0)
  divides 8 2
true

(*divides x*) is **true** when applied
to a number which divides *x*.

  divides 8 3
false
  h(1, 2, 3, 4, 5)
1
  t(1, 2, 3, 4, 5)
2, 3, 4, 5
  prefix 0 (1, 2, 3, 4)
0, 1, 2, 3, 4
  0:(1, 2, 3, 4)
0, 1, 2, 3, 4
  null()
true
  null(1, 2, 3)
false
  0:(1:(2:(3:(4:()))))
0, 1, 2, 3, 4
  def *u x* = *x*:()
  def rec *sum x* =
    if *null x*
    then 0
    else (*h x*) + *sum*(*t x*)
  sum(1, 2, 3, 4)
10
The sum of a list of numbers.
  def rec *append x y* =
    if *null x*
    then *y*
    else (*h x*):*append*(*t x*)*y*
  append(1, 2, 3)(4, 5, 6)
1, 2, 3, 4, 5, 6

  prefix(1, 2, 3)(4, 5, 6)
(1, 2, 3), 4, 5, 6
  def rec *reverse x* =
    if *null x*
    then ()
    else *append*(*reverse*(*t x*))(*u x*)
  reverse(1, 2, 3, 4)
4, 3, 2, 1
  def rec *map f x* =
    if *null x*
    then ()
    else (*f*(*h x*)):*map f*(*t x*)
  map square(1, 2, 3, 4)
1, 4, 9, 16
  map length
    ((), *u* 1, (1, 2), (1, 2, 3))
0, 1, 2, 3
  def rec *concat x* =
    if *null x*
    then ()
    else *append*(*h x*)(*concat*(*t x*))
  concat((1, 2), (3, 4), (5, 6))
1, 2, 3, 4, 5, 6
  def *postfix x y* =
    append y(*u x*)
  postfix 7(5, 6, 8)
5, 6, 8, 7
  def rec *compose f x* =
    if *null f*
    then *x*
    else *compose*(*t f*)(*h f x*)
  compose(*plus* 3,*mult* 2)5
16
  def rec *sumls x* =
    if *atomic x*
    then *x*
    else *sum*(*map sumls x*)

  sumls(1, (2, 3), 4)

10
   def rec *mapls f x* =
      if *atomic x*
      then *f x*
      else *map* (*mapls f*)*x*
mapls square (1, (2, (3, 4), 5))
(1, (4, (9, 16), 25))
   def rec *revls x* =
      if *atomic x*
      then *x*
      else *reverse*(*map revls x*)

Reverses all the lists in a list structure.

   revls(1, (2, (3, 4), 5))
((5, (4, 3), 2), 1)
   reverse (1, (2, (3, 4), 5))
((2, (3, 4), 5), 1)
   def rec *zip x y* =
      if *null x*
      then ()
      else (*h x, h y*):*zip*(*t x*)(*t y*)
   def rec *sp x y* =
      if *null x*
      then 0
      else (*h x* × *h y*) + *sp*(*t x*)(*t y*)
   sp(1, 2, 3)(4, 5, 6)
32
   def *unzip x* = *map* 1*st x, map* 2*nd x*
   zip(1, 2, 3)(4, 5, 6)
(1, 4), (2, 5), (3, 6)
   unzip((1, 6), (2, 5), (3, 4))
(1, 2, 3), (6, 5, 4)
   def *and x y* =
      if *x*
      then *y*
      else false
   def *or x y* =
      if *x*
      then true
      else *y*

def *not x* =
   if *x*
   then false
   else true
def rec *exists p x* =
   if *null x*
   then false
   else if *p*(*h x*)
      then true
      else *exists p*(*t x*)
def *equal x y* = *x* = *y*
 exists (*equal* 5)(2, 6, 1, 5, 7)
true
   def rec *all p x* =
      if *null x*
      then true
      else if *p*(*h x*)
         then *all p*(*t x*)
         else false
 all(*greater* 5)(7, 8, 6, 10)
true
   def rec *filter p x* =
      if *null x*
      then ()
      else if *p*(*h x*)
         then (*h x*):*filter p*(*t x*)
         else *filter p*(*t x*)
 filter odd (1, 4, 6, 5, 8, 7, 2)
   where *odd x* =
      second(*div x* 2) = 1
1, 5, 7

def *belongs x* = *exists*(*equal x*)
def *incl x y* = *all*(*belongs x*)*y*
def *equalset x y* =
   and(*incl x y*)(*incl y x*)
def *intersection* = *filter•belongs*
 intersection(1, 2, 3, 4, 5)
          (3, 4, 5, 6, 7)
 3, 4, 5

def *difference* =
     filter·(not·belongs)

difference(1, 3, 5, 7, 9)(1, 2, 3, 4)
2, 4

def *union x y* =
     append(difference y x)y

def *related x y z* =
     if $x = z$
     then *y*
     else ()

def *brother* = *related joe(al, bob)*

brother joe
al, bob
brother fred
()

def *phi f a b x* = $f(a\ x)(b\ x)$

def *orreln* = *phi union*

def *sister* = *related joe (anne, jean)*

orreln brother sister joe
al, bob, anne, jean

def rec *sumset x* =
     if *null x*
     then ()
     else *union(h x)(sumset(t x))*

def *prodf f g x* =
     sumset (map g(f x))

def *grandfather* =
     prodf pa(orreln pa ma)

def *map2 f x y* =
     let *g z* = *map(f z)y*
        concat(map g x)

def *crossproduct* = *map2 pair*

crossproduct(1, 2, 3)(4, 5, 6)
(1, 4), (1, 5), (1, 6),
(2, 4), (2, 5), (2, 6),
(3, 4), (3, 5), (3, 6)

1.7   TRANSLATING TO A SIMPLER FORM

Section 1.5 contained many additional forms which expressions could take. The details of these will be summarized in a translator from this extended notation to the simple form in which expressions are constructed by application and abstraction alone. The basic syntax of combinations and lambda expressions has been extended by adding the following constructions.

1. Infixed operators
2. Conditional expressions
3. Expressions qualified by definitions
4. List structures in bound variable positions
5. Lists of expressions
6. Definitions of functions
7. Circular definitions
8. Simultaneous definitions

The extended expressions have the following structure formed by combining all the additions described in Section 1.5.

- **rec** an expression is:

    simple and is either

    an *identifier*

    or a *lambda expression*

    and has a *bv* which is a bound variable

    and a *body* which is an expression;

    or *compound* and either is

    regular and has a

    rator and a *rand,* both of which are expressions,

    or is *infixed* and has a *rator*

    which is an identifier,

    and a *left* and a *right,*

    both of which are expressions;

    or is a *conditional* and has

    a *condition* which is an expression

    and either is *two-armed*

    and has a *leftarm* and a *rightarm*

    both of which are expressions,

    or is *one-armed*

    and has an *arm* which is an expression;

    or is a *listing* and is an expression-list;

    or is *qualified* and has

    a *main* which is an expression,

    and an *auxiliary* which is a definition;

    and a definition is:

    a *standard definiiton* and has

    a *definee* which is a bound variable

    and a *definiens* which is an expression;

    or is a *function definition* and has

    a *leftside* which is

    a nonnull bound variable-list

and a *rightside* which is an expression;

  or is a *circular definition* and has

    a *body* which is a definition;

  or is *simultaneous* and is

    a definition-list

  or is a *qualified definition* and has

    a *main* which is a definition,

    and an *auxiliary* which is a definition,

where a bound variable is an identifier-list structure.

The function called *trexp,* defined below, translates expressions to the following reduced form.

- An expression is either *simple*

  and is either an *identifier*

    or a *lambda expression,*

      and has a *bv* which is a bound variable

        and a *body* which is an expression,

  or *compound* and has a *rator* and a *rand* both of which are expressions.

The constructors for these expressions will be called *conslambda* for lambda expressions, and *combine* for compound expressions. Some auxiliary functions are defined first.

1. The function *conslist* is a function for transforming an expression-list into an expression.

     def rec *conslist* x =

        if *null* x

        then '()'

        else *combine(combine 'prefix' (h x))(conslist (t x))*

      conslist('a', 'b', 'c') = *'prefix a(prefix b(prefix c()))'*

2. *Consblock* produces the expression $(\lambda x.M)N$ from two expressions M and N, and a bound variable x.

     def *consblock*$(x, (y, z))$ = *combine(conslambda y x)z*

3. *Cons Y* produces $Y(\lambda x.M)$ from x and M.

     def *cons Y*(x, y) = *combine 'Y' (conslambda x y)*

4. *Conscond* produces the expression *(if p)(λ().a, λ().b)()* from *p, a,* and *b.*

def *conscond(p, x)* =
        combine(combine(combine 'if' p)
                        (conslist(map delay x)))
             (conslist())
            where *delay x* = *conslambda(conslist())x*

5. *Translam* converts *λx y z.M,* for example, to *λx.λy.λz.N,* where *N* is the result of translating *M.*

def rec *translam(x, y)* =
    if *null x*
    then *trexp y*
    else *conslambda(h x)(translam(t x, y))*

Now the function *trexp* for transforming expressions can be defined as follows.

def rec *trexp x* =
if *simple x*
then if *identifier x*
    then *x*
    else *translam(bv x, body x)*
else if *regular x*
    then *combine(trexp(rator x))(trexp(rand x))*
    else   if *infixed x*
        then *combine(rator x)*
                (conslist(trexp(left x), trexp(right x)))
        else if *listing x*
            then *conslist(map trexp x)*
            else if *conditional x*
                then *conscond*
                            (trexp(condition x))
                      (trexp (if twoarmed x
                              then *rightarm x*
                              else *"undefined"))*
                    else *consblock(trexp(main x))*

where rec *trdef x* =
 if *standard x*
 then *definee x, trexp (definiens x)*
 else **if** *function definition x*
   then *h(leftside x),*
      translam(t(leftside x), rightside x)
   else if *circular x*
     then let *b, r* = *trdef (body x)*
       b, cons Y (b, r)
     else if *simultaneous x*
       then *unzip(map trdef x)*
       else let *b, r* = *trdef (main x)*
         b, consblock (r, trdef (auxiliary x))

The function *trdef* converts a definition to standard form. The resulting expressions still have list structures of identifiers as their bound variable parts. The method suggested by examples in Section 1.5.1 will be used in order to reduce the structure even further to lambda expressions having a single identifier in bound-variable position. A list structured bound variable will be replaced by a single new identifier, and the identifiers that occur in the body of the lambda expressions will be replaced by the application of a list structure selector to this new identifier.

  The function *trbvs,* defined below, uses a list of identifier list structures in order to obtain these selector names and the new identifiers. New identifiers created by a function called *cid* which operates on an integer and produces an identifier that cannot occur in the original expression. The value of *(cid n)* will be written x_n. The function also uses the function *positionlist,* defined below, which obtains a list of integers that represents the position of an identifier in a list structure-list. This position has then to be translated, by a function called *sels,* to an expression that describes the selecting function. The function *sels* translates the list 2, 4, 1, 3, for instance, to 3rd·1st·4th·2nd.

def rec *trbvs E x* =
 if *identifier x*
 then let *b, p* = *positionlist E x* 1
   if *b*
   then (**if** *null(t p)*

> $\quad$ **then I**
> $\qquad$ **else** *combine*(*sels*(*t p*))(*cid*(*h p*)))
> $\quad$ **else** *x*
> **else if** *lambda expression x*
> $\quad$ **then let** $E1 = postfix(bv\ x)E$
> $\qquad\qquad conslambda(cid(length\ E1))(trbvs\ E1(body\ x))$
> $\quad$ **else** *combine*(*trbvs E*(*rator x*))(*trbvs E*(*rand x*))

def rec *positionlist x y n =*
> $\quad$ **if** *null y*
> $\quad$ **then false,** ()
> $\quad$ **else let** *b, m = positionls x (h y)* 1
> $\qquad$ **if** *b*
> $\qquad$ **then true,** *n:m*
> $\qquad$ **else** *positionlist x (t y)(n + 1)*

and *positionls x y n =*
> $\quad$ **if** *atomic y*
> $\quad$ **then** *x = y,* ()
> $\quad$ **else** *positionlist x y n*

The expression

$$\lambda(w, (x, y), z).f\ x\ w(\lambda s.s\ g\ z\ x)$$

for example, will be translated by (*trbvs*()) to:

$$\lambda x_1.f(\mathit{first} \cdot \mathit{second}\ x_1)(\mathit{first}\ x_1)(\lambda x_2.x_2\ g(\mathit{third}\ x_1)(\mathit{first}\ x_1))$$

If an identifier is found in the list E that is current when it is searched then the identifier is said to occur *bound* in the whole expression, and will be replaced by a new identifier, or by the application of a list of selectors to a new identifier. If an identifier is not found in the list E, then it is said to occur *free* in the whole expression. The identifiers f and g occur free in the expression above, all the others occur bound.

$\qquad$ The resulting expressions have three formats which can be defined as follows

■ $\quad$ An expression is
$\qquad$ an *identifier*
$\qquad$ or it is a *lambda expression* and has
$\qquad\qquad$ a *bound variable* which is an identifier

and a *body* which is an expression
or it is a *combination* and has an *operator* and an *operand*
which are both expressions.

The reduction to this simple structure has been accomplished at the expense of introducing the constants *Y*, *if*, *prefix*, (), and the list selectors: *first*, *second*, *third*, etc. The infixed point denoting functional composition can be removed in favor of the constant *B*, where $B f g x = f(g x)$.

1.8 LAMBDA CONVERSION

The expressions whose structure is given above are the well-formed formulas of the calculus of lambda conversion. The occurrences of identifiers in positions other than as a bound variable can be classified as free or bound occurrences by the following rules.

1. The identifier x occurs free in the expression x.

2. All occurrences of x in $\lambda x.M$ are bound. An occurrence of an identifier y, other than x in $\lambda x.M$, is free or bound according to whether it is free or bound in M.

3. An occurrence of an identifier in the F or A part of $(F\ A)$ is free or bound in $(F\ A)$ according to whether it is free or bound in F or A.

The free/bound variables of an expression are those which have at least one free/bound occurrence in the expression. It is the free variables of an expression that must be given a value in order to find the value of all expressions in which they are contained.

The expressions have important equivalence rules that are valuable for practical programming purposes. The rules determine when one piece of program can be replaced by another without affecting the outcome. These rules are at the basis of the calculus of lambda conversion, and only depend on the structure of the expression. They can therefore be applied in a mechanical fashion without having to know the meaning of an expression. The value of an expression depends only on the values of its subexpressions and not on any other properties. In other words, if the equivalence relation of having the same value is symbolized by $=$, then:

$$\begin{array}{lll} \text{if } M = N & \text{then} & \lambda x.M = \lambda x.N \\ \text{if } F = G & \text{then} & (F\ A) = (G\ A) \\ \text{if } A = B & \text{then} & (F\ A) = (F\ B) \end{array}$$

The rules of substitution which convert expressions into expressions having the same meaning can be expressed in terms of a substitution operator written $[N/x]M$ which will stand for the result of substituting the expression N for free occurrences of x throughout the expression M.

There are three main rules for transforming expressions into equivalent ones. The first is:

(α) $\lambda x.M = \lambda y.[y/x]M$, provided that y does not occur in M.

This rule expresses the fact that the actual choice of bound variable is irrelevant to the meaning of a lambda expression. If the bound variable and all free occurrences of it in the body are changed then the expression has the same value. This rule needs some qualification; the new identifier must not occur free in the expression or else a free occurrence would become a bound occurrence.

The function *subst* for substituting an identifier x for all free occurrences of the identifier y in the expression z is defined below.

> **def rec** *subst x y z* =
>> **if** *identifier z*
>> **then if** $y = z$
>>> **then** x
>>> **else** z
>> **else if** *lambda expression z*
>>> **then if** $y = bv\ z$
>>>> **then** z
>>>> **else** *conslambda*($bv\ z$) (*subst x y*(*body z*))
>>> **else** *combine*(*subst x y*(*rator z*))
>>>> (*subst x y*(*rand z*))

The second rule is applicable to an expression that is qualified by an auxiliary definition. The expression M **where** $x = N$ must have the same meaning as the result of substituting N for all free occurrences of x within M. This is the second rule of lambda conversion.

(β) $(\lambda x.M)N = [N/x]M$ provided that the bound variables of M are distinct from the free variables of N.

The third rule is the replacement of the right hand side of (β) by its left hand side. If an expression B can be obtained from an expression A by using these rules then A and B are said to be *convertible*, written A *conv* B. A step that replaces the left side of rule (β) by its right side is called a *reduction*, and a step using one application of rule (β) in the other direction is called an *expansion*. If A is convertible to B using only rule (α) and reduction steps then A is said to be *reducible* to B. A reduction represents the process of calculating the results of applying the functions that occur in the expression. An expression that cannot be reduced is said to be in *normal form*. The normal form of an expression is the simplest expression having the same intuitive interpretation. A composite expression whose operator is a lambda expression will be called a *beta-redex*. It follows that an expression in normal form has no beta-redexes.

Successive reductions of an expression, with possible applications of rule (α) provide a method of evaluating an expression and producing its normal form. It may be the case that an expression has no normal form, e.g., the reduction step does not change $(\lambda x.x\ x)(\lambda x.x\ x)$. An example of this reduction to normal form follows.

$$(\lambda f\ g\ x.f(g\ x))(\lambda v\ u.v\ u\ u)(\lambda r\ s\ t.r\ t\ s)$$
$$(\lambda g\ x.(\lambda v\ u.v\ u\ u)(g\ x))(\lambda r\ s\ t.r\ t\ s)$$
$$(\lambda g\ x.\lambda u.g\ x\ u\ u)(\lambda r\ s\ t.r\ t\ s)$$
$$\lambda x.\lambda u.(\lambda r\ s\ t.r\ t\ s)x\ u\ u$$
$$\lambda x.\lambda u.(\lambda s\ t.x\ t\ s)u\ u$$
$$\lambda x.\lambda u.(\lambda t.x\ t\ u)u$$
$$\lambda x.\lambda u.x\ u\ u$$

The (β) rule cannot be applied without some qualification because a free variable of N might become bound when N is substituted for x in M. One way to avoid this is to impose the restriction that the free variables of N are distinct from the bound variables of M. Another way is to change the bound variables of M so that they differ from the free variables of N.

Since an expression may contain more than one beta-redex there is a choice to be made in the reduction step that is to be taken next. The main theorem of the calculus of lambda conversion states (roughly) that this choice makes no difference to the eventual outcome. The Church-Rosser theorem [1–5] states that if A is convertible to B, there is a conversion in which no expansion step precedes any reduction step. It follows that if B is a normal form of A, A is reducible to B, and the normal form is unique [to within applications of rule (α)]. As far as expressions that have a normal form are concerned, the normal form of an expression may be considered to be its value since it can be used to represent the equivalence class of all convertible expressions.

Two different calculi of lambda conversion can be established. In the first, called the λ-**I** calculus a well-formed expression must have at least one free occurrence of the bound variable in its body. In the second, the λ-**K** calculus, there is no such restriction. In the λ-**I** calculus every subexpression of an expression that has a normal form also has a normal form, and any sequence of reductions will produce that normal form in a finite series of steps. This is not true in the λ-**K** calculus, because a subexpression not having a normal form may become cancelled in the reduction steps. If an expression of the λ-**K** calculus has a normal form, then a series of reductions that always reduce an outermost beta-redex first will always produce the normal form because such a series ensures that all such cancellations are made.

The calculus of lambda conversion is a formal system in which any two interconvertible expressions have the same meaning in any interpretation. Although the expressions can be given any interpretation, it is most natural to interpret $(\lambda x.M)$ as a function, and $(\lambda x.M)N$ as the result of the application of a function to its argument. It is surprising that by using this formalism of expressions having no free variables, together with the two rules (α) and (β), we can precisely define the intuitive notion of an *effectively calculable function of positive integers.*

A brief description of the purely syntactical approach to the definition of the effectively calculable functions follows.

Certain formulas can be chosen to stand for the integers

$$\mathbf{Z}_0 = \lambda f\, x.x$$

$$\mathbf{Z}_1 = \lambda f\, x.f\, x$$

$$\mathbf{Z}_2 = \lambda f\, x.f(f\, x)$$

$$\mathbf{Z}_3 = \lambda f\, x.f(f(f\, x))$$

The integer n is identified with the function which when applied to an argument f yields the product of f with itself n times. Thus $(\mathbf{Z}_4\, f)$ is the function that applies f four times to its argument. The arithmetic relies on the fact that the functional composition of f^n and f^m is f^{n+m}. When these functions representing the integers are composed, the arithmetic is performed upon the exponents. Note that the expressions for the integers are in normal form.

A function f of nonnegative integers is said to be λ-*definable* if there is a formula F such that, where n and m are positive integers and N and M are expressions for the integers n and m, when $f\, m = n$, then FM *conv* N, and when the function f has no value for a positive integer m, then FM has no normal form. The successor function, for example, may be λ-defined by the expression

$$\lambda x\, y\, z.y(x\, y\, z)$$

since

$$\lambda x\ y\ z.y(x\ y\ z))\mathbf{Z}_n = (\lambda x\ y\ z.y(x\ y\ z))(\lambda f\ x.f^n\ x)$$
$$= (\lambda y\ z.y((\lambda f\ x.f^n\ x)y\ z))$$
$$= \lambda y\ z.y((\lambda x.y^n\ x)z)$$
$$= \lambda y\ z.(y(y^n\ z))$$
$$= \lambda y\ z.y^{n+1}\ z$$
$$= \mathbf{Z}_{n+1}$$

Addition is λ-defined by

$$\lambda m\ n\ f\ x.m\ f(n\ f\ x)$$

which is the functional product of the m-fold application of f and the n-fold application of the function f to x. Multiplication is λ-defined by:

$$\lambda m.\ \lambda n.\ \lambda f.m(n\ f)$$

which is the m-fold application of the n-fold application of a function f. Exponentiation is defined by application,

$$\lambda m.\lambda n.n\ m$$

since, for example,

$$\mathbf{Z}_3\ \mathbf{Z}_2 f\ x = (\mathbf{Z}_2)^3 f\ x = f^8\ x.$$

That the lambda-definable functions can be identified with the functions of positive integers that can be computed using an algorithm, or program, is known as Church's Thesis. The following is quoted from Church's monograph [1–6].

The notion of a method of effective calculation of the value of a function, or the notion of a function for which a method of calculation exists, is not of uncommon occurrence in connection with mathematical questions, but it is ordinarily left on the intuitive level, without attempt at explicit definition. The known theorems concerning λ-definability, or recursiveness, strongly suggest that the notion of an effectively calculable function of positive integers can be given an exact definition by identifying it with that of a λ-definable function or equivalently of a partial recursive function. As in all cases where a formal definition is offered of what was previously an intuitive or empirical idea, no complete proof is possible; but the writer has little doubt of the finality of the identification.

1.9   COMBINATORS

The same subject matter as the lambda-definable functions may be found in the theory of combinators. The fundamental concepts of each theory are definable within the other (with minor qualifications). The theory of combinators does not use the lambda operator but uses expressions built up from sets of elementary functions called *combinators* which embody certain common patterns of application. In this formulation the expressions contain no variables, and each combinator has its own reduction rule. The names of the combinators follow those of Curry and Feys [1–7], in which many other properties of the combinators are discussed. The simplest combinator is the identity function **I**, which leaves its argument unchanged

$$\mathbf{I}\, x = x$$
$$\mathbf{I} = \lambda x.x$$

The combinator **C** reverses the order of the arguments of the function to which it is applied, so that

$$\mathbf{C}\, f\, x\, y = f\, y\, x$$
$$\mathbf{C} = \lambda f\, x\, y. f\, y\, x$$

It follows when (**C I**) is applied to an argument x it produces the function which when applied to a function f, produces $(f\, x)$.

$$\mathbf{C}\,\mathbf{I}\, x\, f = \mathbf{I}\, f\, x = f\, x$$

The combinator **W**, when applied to a function f of two arguments, produces the function of one argument obtained by identifying the two arguments.

$$\mathbf{W}\, f\, x = f\, x\, x$$
$$\mathbf{W} = \lambda f\, x. f\, x\, x$$

(**W** *mult*) is the squaring function and (**W** *plus*) is the doubling function. The combinator for the composition of two functions is called **B**, and is defined as follows:

$$\mathbf{B}\, f\, g\, x = f(g\, x)$$
$$\mathbf{B} = \lambda f\, g\, x. f(g\, x)$$

An alternative notation to **B** $f\, g$ is $f\!\cdot\! g$. A combinator that is disallowed in the λ-**I** calculus is the function for expressing constant functions, **K**:

$$\mathbf{K}\, x\, y = x$$
$$\mathbf{K} = \lambda x\, y.x$$

(**K** x) is the function which produces x when applied to any argument. The three more complex combinators that follow embody certain common patterns of application.

$$\mathbf{S}\,f\,g\,x = f\,x(g\,x)$$
$$\Phi\,f\,a\,b\,x = f(a\,x)(b\,x)$$
$$\Psi\,f\,g\,x\,y = f(g\,x)(g\,y)$$

The predicate $1 < x < 5$, for example could be written without variables as "Φ and(greater 1)(less 5)." and the function for forming the sum of the sines of two numbers as "Ψ plus sin."

The integers may be redefined in terms of the combinators $\mathbf{K}, \mathbf{B}, \mathbf{I}$, and $\mathbf{S}$ as follows:

$$\mathbf{Z}_0 = \mathbf{K}\,\mathbf{I}$$
$$\mathbf{Z}_{n+1} = \mathbf{S}\,\mathbf{B}\,\mathbf{Z}_n$$

Since

$$\mathbf{Z}_0\,f\,x = \mathbf{K}\,\mathbf{I}\,f\,x = \mathbf{I}\,x = x$$

and

$$\mathbf{S}\,\mathbf{B}\,\mathbf{Z}_n\,f\,x = \mathbf{B}\,f\,(\mathbf{Z}_n\,f)x = f(\mathbf{Z}_n\,f\,x) = \mathbf{Z}_{n+1}\,f\,x$$

An ordered pair may be constructed by using the combinator $\mathbf{D}_2$, defined below.

$$\mathbf{D}_2\,x\,y\,z = z(\mathbf{K}\,y)x$$

The first and second of the pair may be selected by applying the pair to $\mathbf{Z}_0$ or $\mathbf{Z}_{n+1}$. For

$$\mathbf{D}_2\,x\,y\,\mathbf{Z}_0 = x$$
$$\mathbf{D}_2\,x\,y\,\mathbf{Z}_{n+1} = y$$

since

$$\mathbf{D}_2\,x\,y\,(\mathbf{K}\,\mathbf{I}) = \mathbf{K}\,\mathbf{I}\,(\mathbf{K}\,y)x = \mathbf{I}\,x = x$$

and

$$\mathbf{D}_2\,x\,y\,\mathbf{Z}_{n+1} = \mathbf{Z}_{n+1}\,(\mathbf{K}\,y)x = \mathbf{K}\,y(\mathbf{Z}_n\,(\mathbf{K}\,y)\,x) = y.$$

The correspondence between the combinator version of ordered pairs and the list notation is as follows:

(x,y)	$\mathbf{D}_2\,x\,y$
$first(x,y)$	$(\mathbf{D}_2\,x\,y)\mathbf{Z}_0$
$second(x,y)$	$(\mathbf{D}_2\,x\,y)\mathbf{Z}_1$

1.10   RECURSIVE FUNCTIONS

The *primitive recursive functions* include all the ordinarily used numerical functions, such as the quotient and remainder in division, the greatest common divisor, the nth prime number, etc. The primitive recursive functions may be defined by using the *combinator of primitive recursion,* **R**, having the properties

$$\mathbf{R}\ a\ g\ \mathbf{Z}_0\ =\ a$$
$$\mathbf{R}\ a\ g\ \mathbf{Z}_{n+1}\ =\ g\ \mathbf{Z}_n\ (\mathbf{R}\ a\ g\ \mathbf{Z}_n).$$

R can be built up from an auxiliary function f,

$$f(s,\ t)\ =\ s\ +\ 1,\ g(s,\ t)$$

or

$$f\,y\ =\ (first\ y)\ +\ 1,\ g(first\ y)(second\ y),$$

which when the combinator versions of successor and ordered pair are used becomes

$$f\,y\ =\ \mathbf{D}_2\ (\mathbf{S}\ \mathbf{B}(y\ \mathbf{Z}_0))(g(y\ \mathbf{Z}_0)(y\ \mathbf{Z}_1)).$$

Now **R** is defined as the second member of the pair produced by the n-fold application of f to $(0,\ a)$.

$$\mathbf{R}\ a\ g\ n\ =\ (n\,f(\mathbf{D}_2\ \mathbf{Z}_0\ a))\mathbf{Z}_1$$

There is a second way to define **R**. In this case the predecessor function, **P**, is used rather than the successor function. The predecessor function can be defined as follows:

$$\mathbf{P}\ n\ =\ \text{let}\ h\ x\ =\ \mathbf{D}_2\ (\mathbf{S}\ \mathbf{B}(x\ \mathbf{Z}_0))(x\ \mathbf{Z}_0)$$
$$n\ h\ (\mathbf{D}_2\ \mathbf{Z}_0\ \mathbf{Z}_0)\ \mathbf{Z}_1$$

in which the function $h(x,\ y)\ =\ x\ +\ 1,\ x$ is applied n times to $(0,\ 0)$ giving $(n,\ n-1)$, and then the second of this pair is the result. Note that

$$\mathbf{P}\ \mathbf{Z}_0\ =\ \mathbf{Z}_0$$
$$\mathbf{P}\ \mathbf{Z}_{n+1}\ =\ \mathbf{Z}_n.$$

In this case $\mathbf{R}\ a\ g\ \mathbf{Z}_{n+1}$ is defined in terms of $\mathbf{R}\ a\ g\ \mathbf{Z}_n$. A function h, is first introduced which is such that

$$h\ a\ g\,f\,\mathbf{Z}_0\ =\ a$$
$$h\ a\ g\,f\,\mathbf{Z}_{n+1}\ =\ g\ \mathbf{Z}_n\ (f\,\mathbf{Z}_n),$$

namely

$$h \, a \, g \, f \, w = ((\mathbf{D}_2 \, (\mathbf{K} \, a)(g(\mathbf{P} \, w)))w)(f(\mathbf{P} \, w)).$$

The next step is to define a function **Y** having the property

$$\mathbf{Y} f = f(\mathbf{Y} f)$$

and then to define **R** by

$$\mathbf{R} \, a \, g = \mathbf{Y}(h \, a \, g).$$

The function **Y** (called the *paradoxical combinator*) is defined as follows.

$$\mathbf{Y} f = (\mathbf{W}(\mathbf{B} \, f))(\mathbf{W}(\mathbf{B} \, f))$$
$$= \mathbf{W} \, \mathbf{S}(\mathbf{B} \, \mathbf{W} \, \mathbf{B})f$$

The property $\mathbf{Y} f = f(\mathbf{Y} f)$ can be proved as follows

$$\mathbf{Y} f = \mathbf{W} \, \mathbf{S}(\mathbf{B} \, \mathbf{W} \, \mathbf{B})f$$
$$= \mathbf{S}(\mathbf{B} \, \mathbf{W} \, \mathbf{B})ff$$
$$= \mathbf{B} \, \mathbf{W} \, \mathbf{B} \, f(\mathbf{B} \, \mathbf{W} \, \mathbf{B} \, f)$$
$$= \mathbf{W}(\mathbf{B} \, f)(\mathbf{B} \, \mathbf{W} \, \mathbf{B} \, f)$$
$$= \mathbf{B} \, f(\mathbf{B} \, \mathbf{W} \, \mathbf{B} \, f)(\mathbf{B} \, \mathbf{W} \, \mathbf{B} \, f)$$
$$= f(\mathbf{B} \, \mathbf{W} \, \mathbf{B} \, f(\mathbf{B} \, \mathbf{W} \, \mathbf{B} \, f))$$
$$= f(\mathbf{Y} f)$$

The primitive recursive functions are those defined in terms of other primitive recursive functions using the following forms of definition:

1. $f(x) = successor(x)$
2. $f = 0$
3. $f(x_1, x_2, \ldots, x_k) = x_i$
4. $f(x_1, x_2, \ldots, x_k) = g(h_1 (x_1, x_2, \ldots, x_k),$
   $\qquad\qquad\qquad\qquad h_2 (x_1, x_2, \ldots, x_k),$

   $\qquad\qquad\qquad\qquad\qquad .$

   $\qquad\qquad\qquad\qquad\qquad .$

   $\qquad\qquad\qquad\qquad h_p (x_1, x_2, \ldots, x_k))$
5. $f(x_1, x_2, \ldots, x_{k-1}, 0) = g(x_1, x_2, \ldots, x_{k-1})$
   $f(x_1, \ldots, x_{k-1}, n + 1) = h(x_1, \ldots, x_{k-1}, f(x_1, \ldots, x_{k-1}, n))$

in which $h_1, h_2, \ldots, h_p, h$, and g are primitive recursive functions. They may be lambda-defined by choosing:

1. *successor* = **S B**
2. $0 = \mathbf{Z}_0$
3. $f = \mathbf{K}^{i-1}\,\mathbf{K}^{k-i} = \lambda x_1\,x_2 \dots x_k \cdot x_i$
4. $f = \Phi_{(k,p)}\,g\,h_1\,h_2\,h_3 \dots , h_p$

   where

$$\Phi_{(k,p)}\,g\,h_1\,h_2 \dots h_p\,x_1\,x_2 \dots x_k = g(h_1\,x_1\,x_2 \dots x_k)$$
$$(h_2\,x_1\,x_2 \dots x_k)$$
$$\dots$$
$$\dots$$
$$(h_p\,x_1\,x_2 \dots x_k)$$

5. $fx_1\,x_2 \dots x_{k-1} = \mathbf{R}\,(g\,x_1\,x_2 \dots x_{k-1})(h\,x_1\,x_2 \dots x_{k-1})$

Virtually all the algorithmic functions of ordinary mathematics can be shown to be primitive recursive. It is possible, however, to construct functions that are obviously algorithmic but which are not primitive recursive. It is the *partial recursive functions,* defined by a set of recursion equations, which *are* identified with the effectively calculable functions. It has been shown that their computation can take the form of an unbounded search for an integer satisfying some effective condition. By Kleene's normal form theorem [1–11] all partial recursive functions can be defined in terms of two primitive recursive functions h and g as follows:

$$f(x_1, x_2, \dots, x_n) = h(\mu\,k.g(x_1, x_2, \dots, x_n, k) = 0)$$

in which $\mu k.g(x_1, x_2, \dots, x_n, k) = 0$ is the least value of k, if any, such that

$$g(x_1, x_2, \dots, x_n, k) = 0,$$

and is undefined if no such value exists. Suppose the function μ has the property that if there is an integer $n \geq m$, such that $p\,n = \mathbf{Z}_0$, then $\mu\,p\,m$ is the smallest such integer. The function μ is defined in terms of an auxiliary function f as follows:

$$\mathbf{let}\ f\ x\ y\ z = \mathbf{D}_2\,z(x\ y(\mathbf{S}\ \mathbf{B}\ z))(y\ z)$$
$$\mu = \mathbf{Y}\,f$$

The partial recursive function f may now be defined as:

$$f\,x_1\,x_2 \dots x_n = h(\mu\,(g\,x_1\,x_2 \dots x_n)\,0)$$

1.11 TRANSLATION TO COMBINATORS

The combinators may be removed from an expression by substituting their defining lambda expressions. It is possible to translate in the other direction and eliminate lambdas from an expression at the expense of introducing combinators. This demonstrates that, although variables are a useful device in practice, they are logically unnecessary. There are many ways to carry out this translation. The simplest is to produce combinations of **S** and **K**. The following program operates on an expression E and an identifier x which is to be abstracted from E. It removes all lambda operators by introducing the combinators, **I**, **K** and **S**. The **I** combinator may be eliminated by substituting 'SKK' for its occurrences. The function for constructing a combination will be called *combine*.

> **def rec** *extract* x E =
>> **if** *identifier E*
>> **then if** $E = x$
>>> **then** 'I'
>>> **else** *combine* 'K'E
>> **else if** *lambdaexp E*
>>> **then** *extract* x(*extract*(*bv E*)(*body E*))
>>> **else let** $F = $ *extract* x(*rator E*)
>>>> **let** $A = $ *extract* x(*rand E*)
>>>> *combine* 'S'(*combine F A*)

The expression $\lambda f.\lambda x.f\, x\, x$ would be translated as follows:

> $\lambda f.\lambda x.f\, x\, x$
> $\lambda f.\mathbf{S}(\mathbf{K}\, f)(\mathbf{S\, I\, I})$
> $\mathbf{S}(\mathbf{S}(\mathbf{K\ S})(\mathbf{S}(\mathbf{K\ K})\mathbf{I})))(\mathbf{S}(\mathbf{K\ S})(\mathbf{S}(\mathbf{K\ I})(\mathbf{K\ I})))$

It can be seen that the expression expands rapidly and that there is some room for "optimization" by detecting special cases. The program below detects whether the variable occurs in the *rator* and *rand* of the combination before constructing its equivalent. The special cases of

$$\mathbf{S}f g\, x = (f\, x)(g\, x)$$

that are detected are:

1. If the operator does not depend on x, a **B** is introduced

$$\mathbf{B}f g\, x = (f)(g\, x)$$

2. If the operand does not depend on x, a **C** is introduced.

$$\mathbf{C} f g x = (f x)(g)$$

3. If neither operand nor operator depends on x then no combinator is introduced.

$$f g = (f)(g)$$

4. When $g = \mathbf{I}$

$$\mathbf{S} f \mathbf{I} x = (f x)(\mathbf{I} x) = f x x = \mathbf{W} f x$$
$$\mathbf{B} f \mathbf{I} x = (f)(\mathbf{I} x) = f x$$

5. When $f = \mathbf{K}$

$$\mathbf{S} \mathbf{K} g x = (\mathbf{K} x)(g x) = \mathbf{I} x$$
$$\mathbf{C} \mathbf{K} g x = (\mathbf{K} x)(g) = \mathbf{I} x$$

The program is

```
def rec extract x E =
    if identifier E
    then if E = x
         then true, 'I'
         else false, E
    else if lambdaexp E
        then let b1, E1 = extract(bv E)(body E)
             let E2 = if b1
                      then E1
                      else combine 'K' E1
             extract x E2
        else let b1, E1 = extract x(rator E)
             let b2, E2 = extract x(rand E)
             if b1
             then if E1 = 'K'
                  then true, 'I'
                  else if b2
                       then if E2 = 'I'
                            then true, combine 'W' E1
                            else true, combine
```

'S'

(*combine E*1 *E*2)

else *combine* 'C' (*combine E*1 *E*2)

  else if *b*2

    then if *E*2 = 'I'

      then true, *E*1

      else true, *combine* 'B' (*combine E*1 *E*2)

    else false, *combine E*1 *E*2

The function *extract* produces a pair whose first member is a truth value. If **true**, then x is free in E and the second member is the result. If **false** then the result is *combine* 'K' *E*1, where *E*1 is the second member of the pair.

REFERENCES AND BIBLIOGRAPHY

The calculi of lambda conversion were introduced by Church [1–3] and are described in his monograph [1–6]. A parallel study in terms of combinators was developed by Curry, and is published in two volumes [1–7, 1–8]. Introductions to Combinatory Logic may be found in Rosenbloom [1–17]; Feys, and Fitch [1–9]; Böhm, and Gross [1–1, 1–2]; and in Hindley, Lercher, and Seldin [1–12].

Three equivalent notions arose independently at about the same time, namely general recursiveness, lambda-definability and computability. The equivalence of the lambda definable functions with general recursive functions was proved by Church [1–4] and Kleene [1–10]. The equivalence of computable and lambda definable functions was proved by Turing [1–18].

The first programming language that was based on recursive function notation was McCarthy's LISP [1–14, 1–15, 1–16]. The language used here depends heavily upon both LISP and on Landin's ISWIM language [1–13].

1–1. Böhm, C., and W. Gross, "Introduction to the CUCH," in *Automata Theory*, E. R. Cainiello (ed.), New York and London: Academic Press, 1966, pp. 35–65.

1–2. Böhm, C., "The CUCH as a formal description language," in *Formal Language Description Languages for Computer Programming*, T. B. Steel, Jr. (ed.), Amsterdam: North Holland, 1966, pp. 179–197.

1–3. Church, A., "A set of postulates for the foundation of logic," *Ann. of Math.* (2) Vol. 33, 1932, pp. 346–366; Second paper *Ann. of Math.* (2), Vol. 34, 1933, pp. 839–864.

1–4. Church, A., "An unsolvable problem of elementary number theory," *Amer. J. of Math.*, Vol. 58, 1936a, pp. 345–363.

1-5. Church, A., and J. B. Rosser, "Some properties of conversion," *Trans. Amer. Math. Soc.,* Vol. 39, 1936b, pp. 472–482.

1-6. Church, A., "The calculi of lambda conversion," *Ann. of Math. Studies,* Vol. 6, Princeton, N.J.: Princeton University Press, 1941.

1-7. Curry, H. B., and R. Feys, *Combinatory Logic Vol. 1,* Amsterdam: North Holland, 1958.

1-8. Curry, H. B., J. R. Hindley, and J. P. Seldin, *Combinatory Logic Vol. II,* Amsterdam: North Holland, 1972.

1-9. Feys, R., and F. B. Fitch, *Dictionary of Symbols in Mathematical Logic,* Amsterdam: North Holland, 1969.

1-10. Kleene, S. C., "λ–definability and recursiveness," *Duke Math. J.,* Vol. 2, 1936, pp. 340–353.

1-11. Kleene, S. C., *Introduction to Metamathematics,* van Nostrand, Princeton, 1964.

1-12. Hindley, J. R., B. Lercher, and J. P. Seldin, *Introduction to Combinatory Logic,* Cambridge, England: Cambridge University Press, 1972.

1-13. Landin, P. J., "The next 700 programming languages," *CACM,* Vol. 9, No. 3, 1966, pp. 157–164.

1-14. McCarthy, J., "Recursive functions of symbolic expressions and their computation by machine, Part 1, *CACM,* Vol. 3, No. 4, 1960, pp. 184–195.

1-15. McCarthy, J., P. W. Abrahams, D. J. Edwards, T. P. Hart, and M. I. Levin, *LISP 1.5 Programmers Manual,* Cambridge, Mass.: M.I.T. Press, 1962.

1-16. McCarthy, J., "A basis for a mathematical theory of computation," (in) *Computer Programming and Formal Systems,* P. Braffort and D. Hirshberg (eds.), Amsterdam: North Holland, 1963, pp. 33–70.

1-17. Rosenbloom, P. C., *The Elements of Mathematical Logic,* New York: Dover, 1950.

1-18. Turing, A. M., "Computability and λ–definability," *J. Symbolic Logic,* Vol. 2, 1937, pp. 153–163.

2
Program Structure

2.1 INTRODUCTION

The second part of Chapter 1 contained a number of examples of programming using a particular notation of combinatory logic. The main concern of this branch of combinatory logic (called combinatory arithmetic) has been with questions of the existence or nonexistence of computational methods rather than with their efficiency or good program design. At first sight the notation of the lambda calculus does not seem to be a very close model of a programming language, but some syntactical devices to increase its usefulness in programming were given in Chapter 1. Similarly, the method of evaluating an expression by reducing it to normal form does not at first glance appear to be as appealing a computer model as do the simple operations of a Turing machine. In this chapter, alternative methods for evaluating the same expressions, which more closely correspond with current programming practice, are considered. While retaining the basic expression structure the programming system will be extended:

1. by representing the values of expressions such that programs may be written in terms of a set of primitive objects which are more familiar to computer users than are expressions in normal form,

2. by considering more efficient methods of evaluating expressions.

In spite of these changes, the expressions still retain the two properties that make them extremely valuable for practical programming purposes.

1. The value of an expression depends only on the *values* of its subexpressions and on none of their other properties.
2. Two convertible expressions have the same value.

The expressions provide one way of explaining the semantics of programming languages. It is possible to establish correspondences with expressions of all kinds, blocks, procedures, coroutines, and **own** variables. If the object were merely semantic specification, then the problem would be solved by simply detailing these correspondences and relying on the known semantics of the notation of the lambda calculus. It is possible to go further than this, however, and to prescribe the run-time behavior of the computer that corresponds to each structural feature of an expression. This is the main subject of this chapter.

Evaluating machines are built up step by step by adding new features to a machine that interprets reverse Polish notation, and culminate in Landin's *SECD* machine. There are two main types of machine: in the first type the expression itself guides the computation; and in the second type the expression is first compiled to produce a list of instructions or program and the value of the expression is then obtained by executing this program.

There are two important programming notions that cannot be explained by using expressions as a model. The first is the jumping or **go to** instruction. A generalized jumping operation has been added to the *SECD* machine to model **go to**'s and labels. This generalization of the jumping possibilities found in current programming languages provides new features that seem to be useful in practice. The second feature not provided by pure expressions is the assignment statement. Although some cases of local assignment can be treated as auxiliary definitions, more global assignments cannot be described in this way. This is because an assignment statement overwrites the information in a position in the state of a machine simultaneously changing all objects in the state that contain that position. The effect of an assignment statement is explained here by describing the components of the state of the machine and specifying how the components share the state's memory positions.

The evaluation method used throughout most of this chapter evaluates the operand before applying the value of the operator to it. The final section of this chapter contains some alternative strategies in which certain computations are advanced or delayed.

2.2   REVERSE POLISH PROGRAMS

The evaluating machines described in this chapter are elaborations of the familiar technique of translating an arithmetical expression to a reverse Polish program and then executing this program by using a pushdown list for storing the intermediate results of the computation. In Chapter 1, we showed that arithmetical expressions are a special case of expressions in the notation of the lambda calculus. In this chapter, we demonstrate how this compiler and machine can be generalized to accommodate these more general types of expressions.

We start with informal accounts of how a reverse Polish compiler works and how the program produced by it is executed. If the arguments of a function must be evaluated before the function is applied and the order in which arguments must be evaluated is fixed, then there is no choice in the order of the steps to be taken during the evaluation of an expression. If the arguments are always evaluated from left to right, then the expression whose tree is given in Fig. 2.1 can be translated to the reverse Polish program:

$$a\,b\,+\,c\,-\,a\,c\,f\,+.$$

The value of the expression is then found by interpreting this program. In this program, a, b, and c are instructions to obtain the values of a, b, and c, and to add these values to the top of a pushdown list, while $+$, $-$, and f are instructions to apply the operations *plus, minus,* and f to the arguments which are found at the top of a pushdown list. The intermediate results, the values of subexpressions, are all stored on a pushdown list. The relationships between the expression, program, and pushdown list are shown in the diagram in Fig. 2.2. The lines cutting through the tree represent the states of the computation.

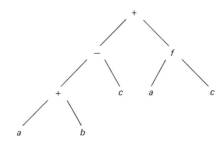

Fig. 2.1   The tree structure of an expression

A line cuts the program at the point which is the position of the instruction-location counter and the tree at points which represent the state of the pushdown list. Each point at which the tree is cut by a line corresponds to a cell of the pushdown list. The value in that cell is the value of the expression whose tree is below that point. Each line cuts the tree into evaluated and unevaluated pieces. The operation in the program that lies between

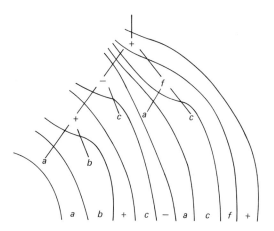

Fig. 2.2   Stages in evaluating an expression

two adjacent lines transforms the pushdown list from the state represented by the line to its left to the state represented by the line to its right. If the values of *a, b,* and *c* are 1, 2, and 3, and $f(x, y) = x^2 + y^2$, then Fig. 2.3 gives the states of the pushdown list during the interpretation of the program.

State	Program	Pushdown list
1	*a*	
2	*b*	1
3	+	1,2
4	*c*	3
5	−	3,3
6	*a*	0
7	*c*	0,1
8	*f*	0,1,3
9	+	0,10
10		10

Fig. 2.3   Program steps and pushdown list

When the end of the program is reached, the value of the expression is on the pushdown list. This is achieved by a program constructed from two types of instruction:

1. loading an object on to the pushdown list,

2. applying an operation to several items at the top of the pushdown list, and replacing them by the result.

To interpret a reverse Polish program the following rules must be observed.

1. Each operation must produce only one result.
2. The same identifier cannot occur in both operator and operand positions.
3. Each operation has a definite number of arguments.

    Suppose a contour is drawn around a set of connected points of the tree as in Fig. 2.4. The part of the tree within the contour will define a new function or subroutine. The contour can be collapsed to a new operation node. The lines pointing into the contour at the bottom represent the argu-

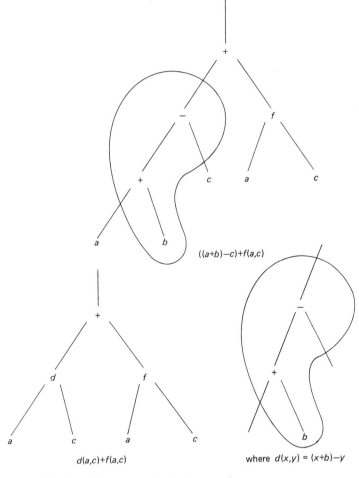

Fig. 2.4   Trees for two equivalent expressions

ments of the function, and the line pointing out of the contour at the top represents the result. Fig. 2.4 shows the situation before and after shrinking the contour. The first tree corresponds to the expression:

$$((a + b) - c) + f(a, c)$$

and the second pair of trees corresponds to the expression:

$$d(a, c) + f(a, c),$$

where

$$d(x, y) = (x + b) - y.$$

The compiler must be defined so that the execution of the two programs that result from these expressions produce the same result. This means that new instructions have to be introduced. The instruction that corresponds to d must cause a subroutine jump; the instructions that correspond to lines crossing the contours, or to occurrences of bound variables, have to be replaced by instructions that obtain the values of a and c. The nature of these instructions and the format of the subroutines will be examined in the following sections.

2.3   VALUES OF IDENTIFIERS

Suppose that there is a mapping E from identifiers to their values which is used to obtain the values of identifiers in an expression. The identifiers are the constants of the system; their values are the primitives. If the value of an identifier x is $(E\ x)$, then a new identifier-value pair can be added to the mapping by the function called *extend:*

> **def** *extend* $(x, y)\ E\ z =$
> > **if** $x = z$
> > **then** y
> > **else** $E\ z$

When a new pair is added to this function, any previous definition of x is effectively overridden. The function E can be represented as a table, or environment, which can either be a list of pairs or a pair of lists. If a list of pairs is used the value can be obtained by the function that follows:

> **def rec** *lookup* $E\ x =$
> > **if** *null E*
> > **then false**, ()
> > **else if** $x = 1st(h\ E)$
> > > **then true**, $2nd(h\ E)$
> > > **else** *lookup* $(t\ E)x$

$$8\,5\,8\,8\,5$$

The function *lookup* returns a pair whose first is **true** if the identifier is found in the list, and its value is the second of the pair.

If the function E is represented as a pair of lists in which the value of an identifier occupies the same position in one list as its identifier does in the other, then *lookup* is redefined as follows:

> **def** *lookup E x* =
>> **let** *b, v* = *position* (1st *E*)*x*
>>> **if** *b*
>>> **then** *select v*(2nd *E*)
>>> **else false**, ()
>>
>> **where rec** *select n x* =
>>> **if** *null x*
>>> **then false**, ()
>>> **else if** $n = 1$
>>>> **then true**, *h x*
>>>> **else** *select* $(n - 1)(t\ x)$

In this second definition the value can be obtained in two stages. First the position is found, and second the value in the same position in the corresponding list is obtained. The two lists can be generalized to list structures by using the function *positionlist* defined in Section 1.8 instead of position. *Positionlist* produces a list of integers rather than a single integer. The function *select* must be replaced by *selectls,* defined as:

> **def rec** *selectls n x* =
>> **let** *b, v* = *select*(*h n*) *x*
>> **if** *b*
>> **then** *selectls*(*t n*)*v*
>> **else false**, ()

To simplify matters, the possibility of failure in this identifier-value function will be ignored, and it will be assumed that every identifier whose value is needed appears in the list. The mapping from identifier to value is realized by a function called *location*, which is such that (*location E x*) is the selector which selects the value of the identifier x from the table E.

2.4   EVALUATING COMBINATIONS

The shape of the program produced by the reverse Polish compiler depends only on the structure of the expression. It is clear that the identifiers could describe any object that can be represented inside a computer or any function that can be carried out by the computer. A more precise description of a similar mechanism, which shows how the program depends on the structure of the expression, will be considered next. The expressions to be evaluated are symmetrical in operator/operand structure, are called *combinations,* and are defined as follows:

> A combination either
>> is an *identifier*
>> or is *compound*
>>> and has an *operator* which is a combination
>>> and an *operand* which is a combination.

The selector names *operator* and *operand* will be abbreviated to *rator* and *rand.* The machine described below finds the value of a combination with respect to a given environment and implements the following function:

> **def rec** *value E x =*
>> **if** *identifier x*
>> **then** *location E x E*
>> **else** (*value E* (*rator x*))(*value E* (*rand x*))

As before, the state of the machine has two components:

1. a *stack,* which is a list of objects, and is used for working space, and
2. a *control string,* which is a list of combinations and represents the program of the machine.

A state whose stack is S and whose control string is C will be written (S, C). The combinations that appear in the control string include a special combination '*apply*' which is used as an instruction to initiate the application of a function to its argument. The definition of the interpreter for combinations is presented below as a function called *transition,* which transforms a state to its successor.

> **def** *transition E* (*S, C*) =
>> **if** *null C*
>> **then** (*S, C*)
>> **else let** $X = h\ C$
>>> **if** *identifier X*
>>> **then** (*location E X E*:*S, t C*)
>>> **else if** *compound X*
>>>> **then** (*S, rand X*:(*rator X*:('*apply*':(*t C*))))

$$\mathbf{else\ if}\ X\ =\ \text{'}apply\text{'}$$
$$\mathbf{then\ let}\ f{:}y{:}S1\ =\ S$$
$$(f\,y{:}S1,\,t\ C)$$

If the machine starts with an empty stack and a control string containing a combination x then, when the control string is exhausted, the repeated application of the transition function will produce the value of x with respect to E on the stack. In other words, the state is transformed from

$$(0, x{:}0)\qquad \text{to}\qquad (value\ E\ x{:}0,\ 0).$$

The machine will interpret any combination containing identifiers defined by E and any compound combination in which the value of the operator part is a function that is applicable to the value of the operand part.

When an identifier is found at the head of the control string, the machine loads its value onto the stack and beheads the control string. When a compound combination is found at the head of the control string, the control string is rearranged. The combination is replaced by a 3-list made up of its operand, its operator, and the special marker '*apply*.' When, in turn, '*apply*' is found at the head of the control, the function found at the head of the stack is applied to its argument (i.e., the second item in the stack), and then both are removed and the result of this application is loaded onto the stack. Again, the control string is replaced by its tail.

The machine merely deals with the structural aspects of the evaluation. The set of objects, the set of identifiers, the identifier-value mapping, and the definition of application are left as parameters. Given these, the machine extends the identifier-value mapping to a combination-value mapping.

The expression $a + b - c + f(a, c)$, for example, may be rewritten as the following combination:

$$p(m(p\ a\ b)c)(f\ a\ c)$$

where

$$p\ x\ y\ =\ x + y$$
$$m\ x\ y\ =\ x - y$$
$$f\ x\ y\ =\ x^2 + y^2$$

The behavior of the machine when it evaluates this combination is illustrated by the sequence of states in Fig. 2.5. The stack is at the left of the page with its head at the right. The control is at the right of the page with its head at the left. The letter 'A' has been used to stand for '*apply*.' The environment contains $a = 1$, $b = 2$, and $c = 3$, and '$p\ 2$,' for example, stands for the function for adding 2.

S	C
()	p(f a c)(m c(p b a))
()	m c(p b a),p(f a c),A
()	p b a,m c,A,p(f a c),A
()	a,p b,A,m c,A,p(f a c),A
1	p b,A,m c,A,p(f a c),A
1	b,p,A,A,m c,A,p(f a c),A
1,2	p,A,A,m c,A,p(f a c),A
1,2,p	A,A,m c,A,p(f a c),A
1,p 2	A,m c,A,p(f a c),A
3	m c,A,p(f a c),A
3	c,m,A,A,p(f a c),A
3,3	m,A,A,p(f a c),A
3,3,m	A,A,p(f a c),A
3,m 3	A,p(f a c),A
0	p(f a c),A
0	f a c,p,A,A
0	c,f a,A,p,A,A
0,3	f a,A,p,A,A
0,3	a,f,A,A,p,A,A
0,3,1	f,A,A,p,A,A
0,3,1,f	A,A,p,A,A
0,3,f 1	A,p,A,A
0,10	p,A,A
0,10,p	A,A
0,p 10	A
10	

Fig. 2.5   Steps in evaluating a combination

There are some fairly obvious shortcuts in which some steps are omitted. These will be discussed in a later section. The machine differs from that described in Section 2.2 in the following respects.

- The intermediate results that are loaded to the stack include the value of every operator.

- Each function has only one argument; if a function has more than one argument, they must be assembled into a list.

- The value of a combination may be a function that is to be applied later.

2.5   A COMPILER FOR COMBINATIONS

A two-stage procedure for evaluating combinations is described next. A combination is first compiled to produce a list of instructions, or program, which is then interpreted to obtain the value of the combination. The compiler, called *revpol*, operates on a combination and an environment and produces a list of instructions of the following type.

■ An instruction either

> is a *load instruction*
>
> > and has an *operand* which is a selector,
>
> or is an *apply* instruction.

The function that produces a load instruction from a selector will be called *consload.* The compiler is defined as follows:

> **def rec** *revpol E x =*
>
> > **if** *identifier x*
> >
> > **then** *consload (location E x)*
> >
> > **else** *concatenate(revpol E (rand x), revpol E (rator x), u'apply')*

The compiler replaces identifiers by their environment positions and rear-ranges compound combinations into lists of three instructions. When these three instructions are interpreted they have the effect of first loading the value of the operand on to the stack, then loading the value of the operator, and finally applying the latter to the former. The transition function for programs, defined below, is simpler than the previous one because the iden-tifier-value mapping and the rearrangement of compound combinations has already been carried out in the compiling stage.

> **def** *transition E* (*S, C*) =
>
> > **if** *null C*
> >
> > **then** (*S, C*)
> >
> > **else let** $X = h \, C$
> >
> > > **if** *load X*
> > >
> > > **then** ($X{:}E{:}S, t \, C$)
> > >
> > > **else if** $X =$ *'apply'*
> > >
> > > > **then let** $f{:}y{:}S1 = S$
> > > >
> > > > ($fy{:}S1, t \, C$)

The revpol function produces the following program from the expression $p(m(p \, a \, b)c)(f \, a \, c)$. In this program the instruction 'apply' has been abbre-viated to *A,* and all the other instructions are *load* instructions.

$$3, 1, f, A, A, 3, 2, 1, p, A, A, m, A, A, p, A, A$$

The expression corresponds to the tree in Fig. 2.6.

This reverse Polish program conforms to the rules given in Section 2.2. There is only one operator *A,* which is never used in the operand position

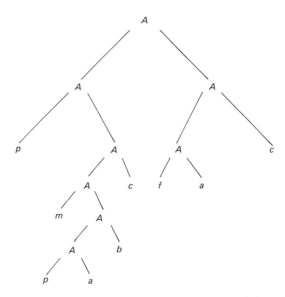

Fig. 2.6 The tree form of the expression $p(m(p\ a\ b)c)(f\ a\ c)$

and always has exactly two arguments. The program string is produced by concatenating the results of compiling the right and left subtrees and *A*, in that order. Figure 2.7 shows the sequence of states during the execution of the program. The sequence is shorter than that for the interpreter, because the transformation from a compound combination to a 3-list of instructions has already been performed in the compiling phase.

S	C
3	3
3,1	1
3,1,*f*	*f*
3,*f* 1	*A*
10	*A*
10,3	*c*
10,3,2	*b*
10,3,2,1	*a*
10,3,2,1,*p*	*p*
10,3,2,*p* 1	*A*
10,3,3	*A*
10,3,3,*m*	*m*
10,3,*m* 3	*A*
10,0	*A*
10,0,*p*	*p*
10,*p* 0	*A*
10	*A*

Fig. 2.7 Steps in running the result of revpol

For combinations, the compiler is no more efficient than the interpreter; the same steps are merely carried out in a different order. The compiler version has been introduced to prepare the way for a more efficient method of evaluating expressions that contain lambda expressions. In this case, the body of a lambda expression may have to be evaluated more than once during the evaluation of an expression containing it; whereas it need only be compiled once. The next section contains some features of programming languages that correspond to expressions containing lambda expressions.

2.6   BLOCKS AND PROCEDURES

The two machines described above may be used to evaluate expressions containing constants only. They will next be extended so that they are capable of evaluating expressions containing variables. These extended machines serve as a model for the implementation of blocks and procedures in a programming language. A brief description follows of the types of implementation problems that arise when a programming language contains blocks and procedures. The examples are taken from ALGOL 60. In ALGOL 60 a block has the syntax

$$\textbf{begin } D; D; D; \dots; S; S; S; S \textbf{ end}$$

in which the D's are declarations, and the S's are statements. A declaration such as **integer** x introduces storage space for an integer which may be referred to in the statements of the block by the name x. The list of declarations in the text will cause a list of storage spaces to be allocated by the program. The program that results from the piece of text in Fig. 2.8 (a) takes the form illustrated in Fig. 2.8 (b). In this instance, the storage spaces for x and y, called *local variables* or *locals* of the block, are only needed

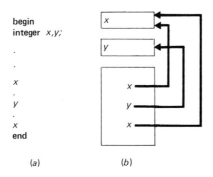

(a)   (b)

Fig. 2.8 An Algol 60 block with a declaration

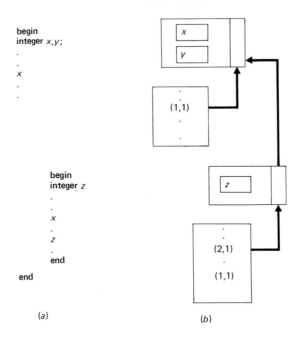

begin
integer x,y;
.
.
x
.
.
.
    begin
    integer z
    .
    .
    x
    .
    z
    .
    end
end

(a)   (b)

Fig. 2.9   Storage for variables in nested blocks

during the execution of the program that corresponds to the block. The blocks may be nested because a statement contained in a block may itself be a block, and the program may take the form illustrated in Fig. 2.9 (a). In this case the variable z is the local of the inner block, whereas x is a local of the outer block, but a nonlocal of the inner block. It is convenient to refer to variables by using two integers. The first is the *depth* of the variable which addresses a list of declarations associated with a block. The locals have *depth* 1, and the locals of the immediately enclosing block have *depth* 2, etc. The second integer refers to the position of the storage for the variable in this list. The translated program takes the form shown in Fig. 2.9 (b).

Since the translation of variables into positions in the environment only depends on the nesting structure of the text, the environment is sometimes called the *static chain.* Four blocks are displayed in Fig. 2.10, together with the corresponding tree of paths through which the variables may be referenced while in each block.

A procedure declaration takes the form:

$$\textbf{procedure } f(x, y, z);$$
$$S$$

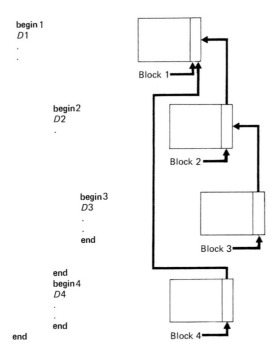

begin 1
D1
.
.

Block 1

begin2
D2
.

Block 2

begin3
D3
.
.
end

Block 3

end
begin4
D4
.
.
end

end

Block 4

Fig. 2.10 Reference to variables

where S is a statement. The (x, y, z) are called *formal variables* and their values are supplied when the procedure is applied to its arguments by a statement or expression such as $f(3, 2, 7)$. The values of other variables in S, however, depend on the position in which the procedure is declared, rather than on the position in which it is applied. In the skeleton of a program in Fig. 2.11 (a) the procedure f is declared in one block and used in an enclosed block.

At the point of applying the procedure f to $(3, 2, 7)$, both the current environment and the point of program execution must be changed. The environment has to be changed to that of the procedure f. The argument $(3, 2, 7)$ is then added to this to form the environment in which the body of the procedure is to be executed. The new point of program execution is the entry to f. After the procedure has been applied, the execution of the program must be resumed after the point of call, and the environment at that point of call must be reinstalled as the current environment. The part of the state that is reserved for storing this linkage information when a procedure is applied to its arguments is called the *dump*. It must contain a location counter, and an environment. This dump must also contain a

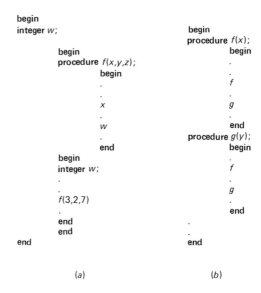

```
begin                                           begin
  integer w;                                      procedure f(x);
                                                    begin
         begin                                        .
           procedure f(x,y,z);                        .
             begin                                    f
               .                                      .
               .                                      g
               x                                      .
               .                                    end
               w                                 procedure g(y);
               .                                   begin
             end                                     .
         begin                                       f
           integer w;                                .
             .                                        g
             .                                        .
           f(3,2,7)                                 end
             .                                     .
         end                                       .
         end                                     end
  end
```

(a)   (b)

Fig. 2.11   Procedure declarations

dump component. Since, if a procedure is called during the evaluation of an expression, it might also deliver a result; the dump must also contain the current stack. The list of states that are chained together by the dump components is often called a *dynamic chain.*

Another feature of ALGOL 60 is that a procedure might have an argument that is a procedure. The collection of information that must be passed must contain both the location of the program for the procedure, and the environment that corresponds to the point at which the procedure is declared.

If several procedures are declared in parallel in the same block, the rules of ALGOL 60 require that the scope of the procedure names includes the procedure bodies. In Fig. 2.11 (b) the procedures f and g are declared in the same block. The occurrences of f and g within the bodies of the procedures refer to the f and g declared in the block rather than to an f or g which might have been declared in enclosing blocks. The two declarations are examples of *mutually recursive* declarations.

On the other hand, the bounds in an array declaration in ALGOL 60 may be defined by an expression. The rule in this case is that the array bound expressions are to be evaluated in the environment of the immediately enclosing block thus disregarding any declarations made in parallel. All questions of the scope of variables, such as these, may be answered by a translation from the program text to the notation of the lambda calculus, in which all scopes of variables are well defined. The corresponding rules for producing programs from the programming language text may then be derived by considering the rules for producing programs from the lambda calculus notation. These rules will be discussed in the next three sections.

2.7 EVALUATING EXPRESSIONS

The problem of evaluating expressions composed by application was the subject of Sections 2.4 and 2.5. In this section, the problem of evaluating expressions composed by abstraction will be considered. We can obtain the interpreter for this problem from the interpreter for combinations by adding two extra actions for 1) loading the value of a lambda expression on to the stack, and 2) applying this value to an argument. These additions amount to choosing a method for representing functions as programs inside the machine. The structural description of the expressions now being considered follows.

An expression is

an *identifier*

or is a *lambda expression*

and has a *bv* which is an identifier

and a *body* which is an expression

or is *compound*

and has a *rator* which is an expression

and a *rand* which is an expression.

The machines for evaluating combinations described in Sections 2.3 and 2.4 both use an environment for assigning values to identifiers. The result of applying a function obtained from a lambda expression is found by evaluating its body. However, the environment must contain the bound variable of the lambda expression, paired with its value, i.e., the argument of the function. Consider the expression

$$ax^2 + bx + c, \qquad \textbf{where} \qquad x = d,$$

or its equivalent expression

$$(\lambda x.ax^2 + bx + c)d.$$

For us to be able to evaluate $ax^2 + bx + c$ the environment must contain the values of a, b, c, and x. However, the value of the lambda expression $\lambda x.ax^2 + bx + c$ does not depend on x. The value of x can only be supplied at the moment when the function is applied to a particular argument, and this argument will be the value of x only during the period that the function is being applied. In the interpreter to be described next, the value of a lambda expression such as $\lambda x.ax^2 + bx + c$ is represented by the lambda expression itself, together with an environment which includes the values of a, b, c, $+$, and square. This collection of information will be called a *closure*. The formation rules for closures are:

> A closure has a *control part* (Cc)
>> which is an expression-list
>> and a *bound variable* (bv)
>> which is an identifier
>> and an *environment part* (Ec)
>> which is an (identifier-object)-pair-list.

A closure will be constructed from its three components by using a function called *consclosure*.

In order to evaluate the expression, $(\lambda x.ax^2 + bx + c)d$, in an environment, $(a = 1, b = 2, c = 3, d = 4)$, the operand d is evaluated first and its value (4) is loaded on to the stack. Then the operator is evaluated, and a closure, which can be written

$$(\text{'}ax^2 + bx + c,\text{'} \text{ 'x,' } (a = 1, b = 2, c = 3, d = 4)),$$

is loaded on to the stack. The closure is then applied to its argument. The application of a closure is performed by pairing its bound variable with the argument, and adding this pair to the environment of the closure. The result is then obtained by evaluating the body of the closure in the environment so produced. It follows that

$$value(a = 1, b = 2, c = 3, d = 4) \text{ '}(\lambda x.ax^2 + bx + c)d' =$$
$$value(a = 1, b = 2, c = 3, d = 4, x = 4)\text{'}ax^2 + bx + c'.$$

The value of an expression can be defined as follows.

> **def rec** *value E x* =
>> **if** *identifier x*
>> **then** *location E x E*
>> **else if** *lambda expression x*
>>> **then** *f*
>>>> **where** *f y* = *value*((*bv x, y*): *E*)(*body x*)
>>> **else** (*value E* (*rator x*))(*value E* (*rand x*))

The value of a lambda expression is a function which, when applied to an argument y, results in the value of the lambda-expression body in an environment made up by extending current environment E by pairing the bound variable of the lambda expression with the argument y. The introduction of lambda expressions means that the environment now changes during evaluation. It will therefore be made a component of the state of the machine.

When a closure is applied it would be possible to delegate the evaluation of the body to another machine. The new machine would be supplied both with the body and with the newly constructed environment and would, after some time, deliver the value. The old machine would then load this result on to its stack and resume its computation. If, however, the same machine is used, the current state at the point at which this new evaluation is initiated will have to be stored, so that the computation may be resumed from that point. The component of the state which is reserved for storing this state is called the dump.

The structure of the state of the machine for evaluating expressions has two extra components. It has a stack and a control string component as before, but also has an environment component and a component for storing states, called a *dump*. With these additions the state has the following structure.

A state has a *stack* (Ss) which is an object-list,

and an *environment* (Es) which is an (identifier-object)-pair-list,

and a *control string* (Cs) which is an expression-list,

and a *dump* (Ds) which is a state.

A state will be written as a 4-list. Two more branches must be added to the transition function for combinations: the first, to load a closure when a lambda expression is encountered at the head of the control string; and the second, to apply a closure to its argument. The four state components are named S, E, C, and D, for Stack, Environment, Control, and Dump, and the machine is called the *SECD* machine [2–10]. A definition of its transition follows.

> **def** *transition*(S, E, C, D) =
>   **if** *null* C
>   **then** (h S:$S1$, $E1$, $C1$, $D1$)
>   **where** $S1$, $E1$, $C1$, $D1$ = D
>   **else let** X = h C
>   **if** *identifier* X

then (*location E X E*:*S, E, t C, D*)
else if *lambda expression X*
   then *consclosure*(*body X, bv, X, E*):*S, E, t C, D*
   else if *X* = '*apply*'
      then let *f*:*y*:*S*1 = *S*
         if *closure f*
         then let *consclosure*(*C*1, *J*, *E*1) = *f*
            (), (*J, y*):*E*1, *u C*1, (*S*1, *E, t C, D*)
         else *f y*:*S*1, *E, t C, D*
      else let *combine*(*F, A*) = *X*
         S, E, A:(*F*:('*apply*':*t C*)), *D*

When a closure is applied, the closure and argument are removed from the stack, '*apply*' is removed from the control string, and the resulting state is stored in the dump. The new stack is the null list. The new environment is formed by pairing the bound variable of the closure with the argument and prefixing it to the environment found in the closure. The new control string is a 1-list containing the body of the closure. After the body has been evaluated, the control string is the null list and the action taken is: 1) to remove the result found at the head of the stack, 2) to reinstall the state found in the dump as the current state, and 3) to load the result to its stack. The total effect of applying a closure to an argument is therefore the same as applying a basic function to its argument, namely, both the function and argument are removed from the stack and replaced by the result. The only state transition that is not a mere rearrangement of the state components is that corresponding to the application of a basic function to its argument.

The application of a closure to its argument is similar to the call of a procedure or subroutine. The action taken when the control string is exhausted corresponds to the exit from a subroutine or procedure. The correspondences become even closer when assignment statements and **go to** statements are added to the model.

It should be noted that that any expression can occur as an immediate component of another in three ways:

1. as the operator of a compound expression,
2. as the operand of a compound expression,
3. as the body of a lambda expression.

If the component is a lambda expression, then the first construction implies that a procedure can be applied to an argument; the second means that a procedure can occur as an argument; and the third means that a procedure can occur as a result of the application of a procedure to its argument. The collection of information that makes up the closure is the information which must be passed to a procedure when it has a procedure as an argument. It is also the information produced when a procedure is constructed as the result of application of a procedure to its argument. The sequence of states in Fig. 2.12 is produced when the expression (*twice square* 3) or

$$((\lambda f.\lambda x.f(f\,x))square\ 3)$$

is evaluated. A closure is enclosed in square brackets.

S	E	C	D
()	()	$(\lambda f.\lambda x.f(f\,x))sq\ 3$	D
()	()	$3,((\lambda f.\lambda x.f(f\,x))sq),A$	D
3	()	$((\lambda f.\lambda x.f(f\,x))sq),A$	D
3	()	$sq,\ \lambda f.\lambda x.f(f\,x),A,A$	D
3,sq	()	$\lambda f.\lambda x.f(f\,x),A,A$	D
$3,sq,[\lambda x.f(f\,x),f,()]$	()	A,A	D
()	$(f{=}sq)$	$\lambda x.f(f\,x)$	$(3,(),A,D)$
$[f(f\,x),x,(f{=}sq)]$	()	()	$(3,(),A,D)$
$3,[f(f\,x),x,(f{=}sq)]$	$(f{=}sq)$	A	D
()	$(x{=}3,f{=}sq)$	$f(f\,x)$	$((),(),(),D)$
()	$(x{=}3,f{=}sq)$	$f\,x,f,A$	$((),(),(),D)$
()	$(x{=}3,f{=}sq)$	x,f,A,f,A	$((),(),(),D)$
3	$(x{=}3,f{=}sq)$	f,A,f,A	$((),(),(),D)$
3,sq	$(x{=}3,f{=}sq)$	A,f,A	$((),(),(),D)$
9	$(x{=}3,f{=}sq)$	f,A	$((),(),(),D)$
9,sq	$(x{=}3,f{=}sq)$	A	$((),(),(),D)$
27	$(x{=}3,f{=}sq)$	()	$((),(),(),D)$
27	()	()	D

Fig. 2.12   Example of operation of the SECD machine

2.8   COMPILING EXPRESSIONS

The specific choice of the identifier that serves as the bound variable of a lambda expression is irrelevant to the meaning or value of the lambda expression (provided confusion of bound variables is avoided). This fact suggests that these identifiers need not play any part in the value of the lambda expression. This section will show how they can be removed in a compiling phase.

There are three types of occurrences of identifiers in a lambda expression $\lambda x.M$. The first type is free in the lambda expression with its values obtained from the environment current when $\lambda x.M$ is evaluated. The second type is bound in the lambda expression but is free in its body, such as the occurrences of x in M that refer to the argument of the lambda expression. The value of such an identifier is the argument of the function and is supplied when the function is applied. The third type is bound in the body

M. These are occurrences of interior bound variables.

When a lambda expression occurs in operator position, it is evaluated and immediately applied; if in the operand position, however, it is evaluated once but the same value may be applied more than once. The environment in which the body of a lambda expression is evaluated is therefore a list whose tail satisfies the demands for the values of free variables and whose head is the argument. The tail of the environment is fixed at the moment of evaluation of $\lambda x.M$; the head changes as the function is applied to different arguments; and the body is evaluated each time. Each environment may be grown by as much as the number of nested lambda expressions appearing within an expression which is evaluated with respect to the environment. Each time a particular *occurrence* of a variable is evaluated, its value must be loaded to the stack. The position in the environment of this object does not vary from one time to another. It can be determined once and for all by examining the text and counting the number of nesting lambda expressions that separate the occurrence from its scope. This number characterizes how its value is to be retrieved from the current environment. For occurrences of the same identifier at different lambda levels the depth varies from occurrence to occurrence.

An occurrence of an identifier (or more generally any subexpression) can be specified by a string of selecting functions composed of *rator, rand* and *body*. The depth of a variable is the number of *body*'s in this string. Consider the three occurrences of x in the expression:

$$\lambda x.(g\ x_1)(\lambda y.(\lambda z.x_2\ y\ z)(f\ x_3)).$$

1. x_1 is selected by *rand · rator · body* and has depth 1.
2. x_2 is selected by *rator² · body · rator · body · rand · body* and has depth 3.
3. x_3 is selected by *rand² · body · rand · body* and has depth 2.

All the bound variables could be replaced by their depths without any loss of information. Two expressions that are convertible under rule (α) will be transformed to the same expression. The expression above, for example, is transformed to:

$$\lambda\ .(g\ 1)(\lambda\ .(\lambda\ .\ 3\ 2\ 1)(f\ 2)).$$

Given a list of identifiers that occur free in an expression, all identifiers may be replaced by depths. The free identifiers are replaced by the depth in the list. The depths of the bound identifiers are formed by adding the length of the list to their depths. The function is:

> **def rec** *replace E x* =
>> **if** *identifier x*
>> **then** *position E x*

> **else if** *lambda expression x*
> **then** *cons* λ(*replace*(*bv x*:*E*)(*body x*))
> **else** *combine*(*replace E*(*rator x*))
> (*replace E*(*rand x*))

The expression above would be transformed to the expression:

$$\lambda .(1\ 3)(\lambda .(\lambda .5\ 4\ 3)(2\ 4))$$

by the function *replace*('*g*', '*f*'). The value of an expression can now be obtained, provided a list of values of the free identifiers is supplied. If an identifier occupies the same position in the list *E* as its value does in the list *F*, then the value can be defined as:

> **def rec** *val F x* =
> **if** *integer x*
> **then** *select x F*
> **else if** *lambda expression x*
> **then** *f*
> **where** *f y* = *val*(*y*:*F*)(*body x*)
> **else** (*val F*(*rator x*))(*val F*(*rand x*))

The value is thereby obtained in two stages.

The two compiling processes that replace variables by their depths and flatten the tree structure of the expression into a list of instructions are now combined. In the following reverse Polish compiler, the identifiers that occur free in the expression being compiled are translated, as before, into instructions that contain the location of the associated objects. Occurrences of bound variables are translated into instructions which select a value from the current environment. The resulting program contains no identifiers and is made up of a list containing four types of instructions, defined below.

- An instruction is
    > a *load instruction*(*load*)
    > and has an *operand* which is a position,
    > or is an *apply instruction*(*appl*),
    > or is a *load position instruction*(*lpos*)
    > and has a *position* which is an integer,
    > or is a *load closure instruction*(*ldcl*)
    > and has a *control* which is an instruction-list.

The *loadposition* instruction and *loadclosure* instructions will be constructed from their components by using the functions *conslpos,* and *consldcl,* respectively. The compiler uses two lists during the translation. The first (E), is the list of identifier-value pairs, and is used to find the values of free variables. The second (F) is a list of identifiers which is used for finding the depths of bound variables. The definition of the compiler follows.

> **def rec** *revpol*$(E, F) x$
>   **if** *identifier x*
>   **then let** $b, y = $ *position F x*
>   **if** b
>   **then** $u(conslpos\ y)$
>   **else** $u(consload(location\ E\ x\ E))$
>   **else if** *lambda expression x*
>   **then** $u(consldcl(revpol(E, bv\ x : F)(body\ x)))$
>   **else** *concatenate*$(revpol(E, F)(rand\ x),$
> $revpol(E, F)(rator\ x),$
> $u\ 'appl')$

The environment and control string of the state have to be changed and a state is now redefined as:

- A state has
  - a *stack* which is an object-list
  - an *environment* which is an object-list
  - a *control string* which is an instruction-list
  - a *dump* which is a state.

The environment is an object-list which holds the values of bound variables. The closure must be redefined and the new closure has the structure:

> A closure has an *environment*(Ec) which is an object-list
>   and a *control*(Cc) which is an instruction-list.

The new transition function is defined below.

> **def** *transition*$(S, E, C, D) = $
>   **if** *null C*
>   **then** $h\ S : S1, E1, C1, D1$ **where** $S1, E1, C1, D1 = D$
>   **else let** $X = h\ C$

if *load X*
then *X E: S, E, t C, D*
else if *loadposition X*
  then *select(position X)E: S, E, t C, D*
  else if *loadclosure X*
    then *consclosure(control X, E): S, E, t C, D*
    else if *X = 'apply'*
      then let *f:y:S1 = S*
      if *closure f*
      then let *consclosure(C1, E1) = f*
        (), y:E1, u C1, (S1, E, t C, D)
      else *f y:S1, E, t C, D*

When interpreted the instructions have the following effects.

1. A *load* instruction loads the object connected to it to the stack.

2. A loadposition (lpos) instruction loads the object that is n places from the top of the current environment to the stack, where n is the depth associated with the instruction.

3. A loadclosure (ldcl) instruction constructs a closure from the control string attached to both the instruction and the current environment, and loads it to the stack.

4. An 'apply' instruction finds either a basic function or a closure at the head of the stack. When a basic function is found, the instruction is applied to the argument of the function (i.e., the second item on the stack), and both are replaced by the result. When a closure is found, the state, except for the closure, its argument, and *'apply,'* is preserved in the dump. The new stack is the null list. The new environment is the environment of the closure; the new control string is the control string of the closure.

5. The only other action taken by the machine occurs when a control string is exhausted. In this case the state found in the dump is reinstalled as the current state, and the result is loaded to the stack of this state.

If the initial state is $((), (), revpol(E, ())x, D)$, then repeated application of the transition function will produce:

$$(value \ E \ x:(), (), (), D)$$

provided the expression has a value.

The reverse Polish translator produces the following program from the expression $((\lambda f.\lambda x.f(f\,x))\,sq\,3)$.

load 3	C1: ldcl C2	C2: lpos 1
load sq		lpos 2
ldcl C1		appl
appl		lpos 2
appl		appl

The machine passes through the sequence of states in Fig. 2.13 when this program is interpreted. A lambda expression corresponds closely to a procedure or subroutine. The closure, treated here as the value of a lambda expression, corresponds to the machine program that results from compiling a procedure. The step for applying a closure corresponds to the implementation of the call of a procedure. The step taken when the control string is exhausted corresponds to the actions taken when leaving a procedure.

3	()	load 3	D
3, sq	()	load sq	D
3, sq, [C1,()]	()	ldcl C1	D
()	sq	appl	(3,(),appl,D)
[C2, sq]	sq	ldcl C2	(3,(),appl,D)
3, [C2, sq]	()	(exit)	D
()	(3, sq)	appl	((),(),(),D)
3	(3, sq)	lpos 1	((),(),(),D)
3, sq	(3, sq)	lpos 2	((),(),(),D)
9	(3, sq)	appl	((),(),(),D)
9, sq	(3, sq)	lpos 2	((),(),(),D)
27	(3, sq)	appl	((),(),(),D)
27	()	(exit)	D

Fig. 2.13   Steps taken when executing a compiled program

The implementation of the procedure-calling mechanism, embodied in the *SECD* machine, has been arranged so that the total effects of applying a basic function and a closure are the same. Both replace the argument and function by the result of their application. It follows that a closure and a basic function are interchangeable, and that one may be used in a context that might normally be regarded as the natural context of the other.

An expression that has a lambda expression in operator position corresponds closely to an ALGOL 60 or PL/I block. The bound variable corresponds to the local variables of the block, and the body of the lambda expression corresponds to the statements of the block. The operand corre-

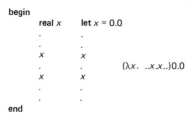

Fig. 2.14 Blocks and **let** expressions

sponds to the initial values given to the locals of the block. In this respect the correspondence breaks down when a declaration such as **integer** x declares a variable x as being an integer but gives no initial value to it. To maintain the correspondence, initial values must be invented and these values will be associated with the variable upon entry to the block. The correspondence between blocks, **let** expressions, and beta-redexes is shown in Fig. 2.14. A lambda expression in operand position corresponds to a declaration of a procedure. The bound variable corresponds to the formal parameters of the procedure, and the body of the lambda expression corresponds to the body of the procedure. An example of this correspondence is illustrated in Fig. 2.15. Some care must be taken in translating the declarations made in parallel in a block. In ALGOL 60 and PL/I the procedures declared in parallel may be mutually recursive and so the scope of a procedure declaration is the body of the block plus the bodies of procedures declared in parallel. The correspondence in this case is with an expression containing an application of **Y** and is illustrated in Fig. 2.16.

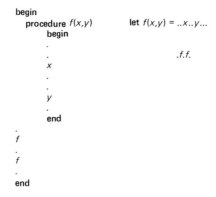

Fig. 2.15 Procedure declarations and auxiliary-function definitions

```
begin
procedure f(x);
        begin       (λ(f,g). ..)(Yλ(f,g).(λx...f.g.,λy..f.g.))
            .
            .
            f
            .
            g
            .
        end
procedure g(y)
        begin
            .
            f
            .
            g
            .
        end
    .
    .
end
```

Fig. 2.16   Mutually recursive procedure declarations

2.9   METHODS FOR INCREASING EFFICIENCY

This section discusses certain structural features of expressions, the detection of which during a compiling phase will increase the efficiency of the resulting program. New instructions are introduced which combine two or more instructions of the machine just described. These composite instructions permit the omission of some actions that would otherwise be necessary.

Lambda expressions as operators.   A lambda expression that occurs in operator position can only be applied once. However, the previous compiler produces the two instructions *ldcl(C)*, and *appl*, in which the *C* part is treated as a subroutine. Alternatively, this program may be implemented by including the control string that results from compiling the body of the lambda expression in line making it unnecessary to either store the control-string part of the state in the dump or load a closure to the stack. Furthermore since the current environment becomes the tail of the new environment after executing these two instructions, it may be recovered from the tail of the new environment rather than from the dump and therefore does not need to be stored in the dump. Two new instructions will be introduced called *enter*, and *exit*, corresponding to the actions to be taken upon block entry and exit. They will bracket the program produced from the body of the lambda expression that occurred in operator position.

The expression $(\lambda x.M)N$ will be translated to

$$program(N)$$
$$enter$$
$$program(M)$$
$$exit$$

rather than

$$program(N) \qquad C:program(M)$$
$$ldcl\,(C)$$
$$appl$$

where *program*(N), for example, is the result of compiling the expression N. The state transformations for the instructions *enter* and *exit* are:

from $(x{:}S, E, enter{:}C, D)$ to $((), x{:}E, C, (S, -, -, D))$

and

from $(r{:}S2, x{:}E, exit{:}C, (S, -, -, D))$ to $(r{:}S, E, C, D)$

The result of compiling an expression is a *descriptive* control string which loads a value on to the stack. This control string may consist of a single instruction or two descriptive control strings (one of which loads the operand and the other the operator) followed by the instruction *appl*. When a control string is prefixed with *enter* and postfixed with *exit* it becomes a *functional* control string that transforms an argument found on the stack. Another way of constructing a functional control string is by postfixing *appl* to a descriptive control string for a function. The block construction is a more efficient method of implementing a particular type of functional control string.

If this feature is added to the compiler the program that results from

$$(\lambda f.\lambda x.f(f\,x))\,sq\,\,3$$

takes the form:

load 3	*C*2: *lpos* 1
load sq	*lpos* 2
enter	*appl*
*ldcl C*2	*lpos* 2
exit	*appl*
appl	

The sequence of states that results from interpreting this program is given in Fig. 2.17.

3	()	*load* 3	D
3,*sq*	()	*load sq*	D
()	*sq*	*enter*	(3,-,-,D)
[C2,*sq*]	*sq*	*ldcl C2*	(3,-,-,D)
3,[C2,*sq*]	()	*exit*	D
()	(3,*sq*)	*appl*	((),(),(),D)
3	(3,*sq*)	*lpos* 1	((),(),(),D)
3,*sq*	(3,*sq*)	*lpos* 2	((),(),(),D)
9	(3,*sq*)	*appl*	((),(),(),D)
9,*sq*	(3,*sq*)	*lpos* 2	((),(),(),D)
27	(3,*sq*)	*appl*	((),(),(),D)
27	()	(*exit*)	D

Fig. 2.17   Steps when executing a block

Variables as operators.   The two instructions *lpos*(p) and *appl* can be combined into a single instruction called *apos*(p), which has their combined effect but in which the function does not need to be loaded to the stack. The program for $C2$, above, becomes:

$$C2:lpos\ 1$$
$$apos\ 2$$
$$apos\ 2$$

Lambda expressions as operands.   A lambda expression which appears in operand position is similar to the declaration of a procedure in ALGOL 60. Each occurrence of its name in the operator lambda expression will give rise, when it is compiled, to a *lpos* instruction which contains the position of the corresponding closure in the environment which becomes the current environment when the instruction is executed. The closure addressed in this way was originally produced by a *ldcl* instruction and incorporates the environment of the state that was current when the closure was constructed. By the time that this closure is invoked by a *lpos*(k) instruction the environment of the closure has become the t^k of the current environment. It is therefore possible to delay the construction of the closure until it is invoked. The two instructions

$$ldcl(C)$$
$$\cdots$$
$$lpos(k)$$
$$\cdots$$

may be replaced by a single instruction which has both the control string C and the position k attached to it. When this new instruction is executed it constructs a closure from C and the t^k of the current environment. Two new instructions will be introduced:

1. *linc(k, C)*, (for load-invoked closure) which loads a closure constructed in this way, and

2. *ainc(k, C)* (for apply-invoked closure) which applies the closure without first constructing it.

The state transformations produced by these two new instructions are:

from $(S,E,linc(k,C1):C,D)$ to $((consclosure(C1,t^k E):S,E,C,D),$

and from $(x:S,E,ainc(k,C1):C,D)$ to $((),x:t^k E,C1,(S,E,C,D)).$

If the squaring function had been implemented as a closure

$$sq = \lambda x.x \times x,$$

then instead of the program

load 3	*C2: lpos* 1	*C3: lpos* 1
*ldcl C*3	*apos* 2	*lpos* 1
enter	*apos* 2	*multiply*
*ldcl C*2		
exit		
appl		

the following program would be produced.

load 3	*C2: lpos* 1
enter	*ainc*(2, *C*3)
*ldcl C*2	*ainc*(2, *C*3)
exit	
appl	

Conditional expressions.   The syntax of a conditional expression can be detected at compile time and a program having the following shape can be produced.

> **if** *a*
> **then** *b*
> **else** *c*
> *program*(*a*)
> *test*(*L*)
> *program*(*b*)
> **go to**(*M*)
> *L*: *program*(*c*)
> *M*: ...

In this program the *test* instruction expects to find a truth value at the head of the stack and removes it. Control is then passed to L if the value is **false**, otherwise the tail of the control string replaces the control. The **go to**(M) instruction changes the control to M. The state transformations of the two new instructions are:

$$\text{from} \quad (\textbf{true}:S, E, \textit{test}(L):C, D) \quad \text{to} \quad (S, E, C, D)$$

and

$$\text{from} \quad (\textbf{false}:S, E, \textit{test}(L):C, D) \quad \text{to} \quad (S, E, L, D)$$

and

$$\text{from} \quad (S, E, \textbf{go to}(M):C, D) \quad \text{to} \quad (S, E, M, D)$$

Self reference and Y. The fixed-point-finding operator **Y** was introduced in chapter one to construct expressions that describe self-referential functions. Methods for implementing this function are considered in this section. One definition of a fixed-point-finding function is

$$\mathbf{Y} = \lambda f.(\lambda g.f(g\,g))(\lambda g.f(g\,g))$$

since by lambda reduction

$$\begin{aligned}
\mathbf{Y}\,F &= (\lambda f.(\lambda g.f(g\,g))(\lambda g.f(g\,g)))F \\
&= (\lambda g.F(g\,g))(\lambda g.F(g\,g)) \\
&= F((\lambda g.F(g\,g))(\lambda g.F(g\,g))) \\
&= F(\mathbf{Y}\,F).
\end{aligned}$$

The version of **Y** that is suitable for the *SECD* machine is

$$\mathbf{Y}f = h\,h \quad \text{where} \quad h\,g = f(\lambda x.g\,g\,x),$$

in which the application of g to itself is delayed until the closure corresponding to $\lambda x.g\,g\,x$ is applied. This avoids infinite looping because in the *SECD* machine the body of a closure is only evaluated when the closure is applied. The application of **Y** as defined above to a function-transforming function such as $(\lambda\,F.\,\lambda x\ldots F(y)\ldots)$ produces a closure which, when applied to an argument x, has the property that the interior application of F applies the same closure to y. This may be shown by tracing through the *SECD* states as follows. Suppose the expression $\mathbf{Y}(\lambda F.\lambda x \ldots F(y)\ldots)x$ is first compiled to produce the program below.

F0: *load x*	F1: *ldcl F2*	F2: *load y*
ldcl F1		*lpos 2*
ldcl Y1		*appl*
appl		
appl		

$Y1$: *ldcl* $Y2$	$Y2$: *ldcl* $Y3$	$Y3$: *lpos* 1
ldcl $Y2$	*lpos* 2	*lpos* 2
appl	*appl*	*lpos* 2
		appl
		appl

When this program is executed in an environment E and with a dump D, it first passes through the states 1 to 14 given in Fig. 2.18. In doing so the closure corresponding to $\mathbf{Y}(\lambda F.\lambda x \ldots F(y) \ldots)$ is constructed, namely:

$$[F2, ([Y3, E2]:E)] \quad \textbf{where} \quad E2 = [Y2, E1]:E1$$

$$\textbf{where} \quad E1 = [F1, E]:E.$$

When this closure is applied to x the state passes through stages 15 to 23. The $S, E,$ and C parts of states 8 and 23 are identical, so the same closure will be constructed for the internal application of F to y. It follows that the *SECD* machine at it stands is capable of implementing functions that contain references to themselves. Systems programmers will be familiar with more straightforward ways for producing programs having loops. Suppose the program text contains both definitions and uses of values, and that when a definition is made an identifier is paired with its value in a table. There are two main ways of dealing with the problem caused by the use of an identifier preceding its definition. In the first method, the positions in which

1. (), E, (*load* x, *ldcl* $F$1, *ldcl* $Y$1, *appl*, *appl*), D
2. x, E, (*ldcl* $F$1, *ldcl* $Y$1, *appl*, *appl*), D
3. ([$F$1, E], x), E, (*ldcl* $Y$1, *appl*, *appl*), D
4. ([$Y$1, E], [$F$1, E], x), E, (*appl*, *appl*), D
5. (), $E1$=([$F$1, E]:E), (*ldcl* $Y$2, *ldcl* $Y$2, *appl*), (x, E, *appl*, D)=$D$1
6. [$Y$2, $E$1], $E$1, (*ldcl* $Y$2, *appl*), $D$1
7. ([$Y$2, $E$1], [$Y$2:$E$1]), $E$1, *appl*, $D$1
8. (), $E2$=([$Y$2, $E$1]:$E$1), (*ldcl* $Y$3, *lpos* 2, *appl*), ((), $E$1, (), $D$1)=$D$2
9. [$Y$3, $E$2], $E$2, (*lpos* 2, *appl*), $D$2
10. ([$F$1, E], [$Y$3:$E$2]), $E$2, (*appl*), $D$2
11. (), ([$Y$3, $E$2]:E), *ldcl* $F$2, ((), $E$2, (), $D$2)=$D$3
12. [$F$2, ([$Y$3:$E$2]:E)], [$Y$3, $E$2]:E, (), $D$3
13. [$F$2, ([$Y$3:$E$2]:E)], $E$2, (), $D$2
14. [$F$2, ([$Y$3, $E$2]:E)], $E$1, (), $D$1
15. ([$F$2, ([$Y$3, $E$2]:E)], x), E, *appl*, D
16. (), (x:[$Y$3, $E$2]:E, (. ., *load* y, *lpos* 2, *appl*, .), ((), E, (), D)
17. y, (x:[$Y$3, $E$2]:E), (*lpos* 2, *appl*, . .), ((), E, (), D)
18. ([$Y$3, $E$2], y), (x:[$Y$3, $E$2]:E), (*appl*, . .), ((), E, (), D)=$D$3
19. (), y:$E$2, (*lpos* 1, *lpos* 2, *appl. appl*), ((), (x:[$Y$3, $E$2]:E), . . . , $D$3)=$D$4
20. y, y:$E$2, (*lpos* 2, *lpos* 2, *appl*, *appl*), $D$4
21. ([$Y$2, $E$1], y), y:$E$2, (*lpos* 2, *appl*, *appl*), $D$4
22. ([$Y$2, $E$1], [$Y$2, $E$1], y), y:$E$2, (*appl*, *appl*), $D$4
23. (), $E$2, (*ldcl* $Y$3, *lpos* 2, *appl*), (y, y:$E$2, *appl*, $D$4)
28. [$F$2, ([$Y$3, $E$2]:E)], $E$2, (), (y, y:$E$2, *appl*, $D$4)

Fig. 2.18 The SECD machine implementation of **Y**

the values are to be placed are chained together so that when the name is defined the positions are filled in. In the second method, the value is referred to indirectly by setting up a dummy piece of storage which will be updated by the value when the value is defined. The problem arises because the definitions are mutually recursive and may be considered to be produced by an application of **Y**. The second method for implementing **Y** is shown below. This version of **Y** will only be applicable to a transformer whose only use of its argument is to either incorporate the argument, or components of the argument, in its result. The result produced must have the same structure as the argument down as far as the components that are incorporated. Let f be a function such that when it is applied to x produces a result y with the same structure as x. Next suppose that w is the result of updating x by resetting each of the components selected by the function to the corresponding components of y. Then if f is applied to w, the result will be w.

When the function-producing function $\lambda f.\lambda z.M,$ is applied to a functional argument x it will produce a closure as a result. This closure will contain the argument x at the head of its environment. The function $\lambda f.\lambda z.M$ is therefore an example of a function which merely incorporates its argument in its result, and both the argument and the result are closures. The action taken by applying **Y** to such a function is to replace the head of the environment of the closure by the closure itself, producing a closure whose structure is given in Fig. 2.19. This can be accomplished in general as follows. First create a dummy argument d which has the same top level structure as the argument and the result of the function **Y** is being applied to. Next apply f to d producing a result y, and update the dummy argument positions of d by the corresponding positions of y. The result of applying **Y** to f is the updated value of d. The structure of the argument and the positions selected must be known in order to implement **Y** in this way. When **Y** is applied to $\lambda f.\lambda z.M$ it is essential to use an indirect representation of the closure. Suppose that the closure is represented by an address which points to a pair of addresses, one for the control string and the other for the environment. Then the dummy argument d will have the shape shown in Fig. 2.19. When $\lambda f.\lambda z.M$ is applied to d it will produce a closure with d at the head of its environment, as shown in Fig. 2.18. The application of **Y** next updates the pair of the dummy by the pair of this resulting closure, producing the closure diagrammed in Fig. 2.18, in which the head of the environment is the closure itself.

This technique is also applicable to functions that transform a list of functions, such as

$$\lambda(f, g, h).((\lambda x \ldots f \ldots g \ldots), (\lambda h \ldots f \ldots h \ldots), (\lambda z \ldots f \ldots g \ldots)).$$

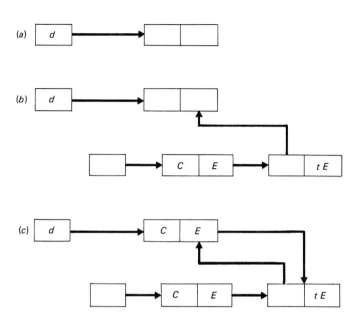

Fig. 2.19 Stages in the application of **Y**

In this case the dummy argument must be a list of length 3 containing three dummy closures as above, and each dummy closure has to be updated by its corresponding closure pair in the same way as above.

Another type of function that is amenable to this treatment is a function which transforms a list structure to a list structure of the same shape. When **Y** is applied, it produces a list structure with loops.

The application of **Y** to a lambda expression is similar to the application of a lambda expression to its argument since

$$\mathbf{Y}(\lambda f.M) = (\lambda f.M)(\mathbf{Y}(\lambda f.M))$$

and the same kind of optimization can be carried out as for lambda expressions in operand position, i.e., where the position is occupied by a closure, replacing $lpos(k)$, by $linc(k, C)$, and $apos(k)$ by $ainc(k, C)$. This produces a closure with a loop in its control string of the form

$$C: \ldots \qquad or \qquad C: \ldots$$

$$\ldots \qquad\qquad\qquad \ldots$$

$$\ldots \qquad\qquad\qquad \ldots$$

$$linc(k, C) \qquad\qquad ainc(k, C)$$

$$\ldots \qquad\qquad\qquad \ldots$$

Collapsing on exit. If the body of a lambda expression is a combination or a conditional expression having an arm which is a combination, then the last thing that happens when it is applied to its argument is that a function is obtained and then applied to an argument. The last instruction in the resulting program is '*appl.*' This is followed by an exit from the program. The sequence of instructions is therefore

$$ldcl\ f$$
$$appl \qquad f\!: \ldots$$
$$(exit) \qquad \ldots$$
$$\qquad\qquad (exit)$$

in which there is an execution sequence *exit, exit*. These two *exit*'s can be collapsed into one by making the first exit encountered look like the second. This may be done by making sure that the first *exit* refers to the dump of the calling program rather than to its own dump, which in fact is never stored. The instructions '*apply*:()' can be replaced by a new instruction called '*applyexit*,' which transforms a state from

$$consclosure(C1,\ E1)\!:\!y\!:\!S,\ E,\ applyexit\!:\!(),\ D \quad \text{to} \quad S,\ y\!:\!E1,\ C1,\ D.$$

Since the only instruction that refers to the dump is the exit, which reinstalls it, this change does not affect the final result. Furthermore, if both closures have the same environment, and *apply* is replaced by *ainc*(1, *C*1), then the exit can be accomplished by a mere change of control string, and a change to the head of the environment, i.e.,

$$\text{from} \qquad y\!:\!S,\ x\!:\!E,\ ainc\,(1,\ C1)\!:\!(),\ D \qquad \text{to} \qquad S,\ y\!:\!E,\ C1,\ D.$$

These conditions are satisfied by the very common recursive calls at depth 1, in which the calling and called functions are the same. The resulting economy is to replace a recursive reentry by a simple **go to** instruction. In this case the two arguments x and y must have the same structure, and if x is not incorporated in y or the called function then x can be updated by y. Consider, for example the *sum*1 function, which adds the numbers in a list x to an initial value a. This is defined as:

$$\textbf{def rec } sum1(a,\ x)\ =$$
$$\qquad \textbf{if } null\ x$$
$$\qquad \textbf{then } a$$
$$\qquad \textbf{else } sum1(a\ +\ h\ x,\ t\ x)$$

and may be implemented by the program

$$sum\,1:\textbf{if } null\ x$$

$$\textbf{then } a$$

$$\textbf{else } (a,\ x): = (a\ +\ h\ x,\ t\ x)$$

$$\textbf{go to } sum\,1$$

Extracting common subexpressions.   The technique of extracting common subexpressions which is used by both programmers and compilers can be considered an example of the expansion rule of lambda conversion. An attempt is made to discover common subexpressions in an expression such as

$$a^2\ +\ b^2\ -\ log(1/(a^2\ +\ b^2))$$

and to ensure that the value of $a^2\ +\ b^2$ is only calculated once. The transformation is an application of the expansion rule of the lambda calculus to produce:

$$x\ -\ log(1/x)\ \text{where } x\ =\ a^2\ +\ b^2.$$

This technique is particularly valuable if it enables calculations to be removed from a loop.

2.10 LABELS AND GO TO STATEMENTS

> Despair thy charm,
> And let the angel whom thou still has served tell thee,
> Macduff was from his mother's womb
> Untimely ripp'd
> *Macbeth, Act v, Sc. viii.*

The **go to** statement introduced in the last section only changes the control string part of the state. When a programming language has a block structure and procedures there are additional jumping possibilities. It is possible in ALGOL 60, for example, to jump from a block to a point in an enclosing block. This kind of jump requires a greater change to the state than just changing control, because the environment corresponding to the block jumped into must be reinstalled. When the jump is made both the control string and the environment have to be changed. An exit from a block by a **go to** instruction will be called an *unnatural exit* as opposed to a *natural exit* which is an exit through the **end.**

In ALGOL 60 it is also possible to pass a label as an argument to a procedure, so that a **go to** instruction inside the procedure body causes an exit from the procedure to the point corresponding to the label argument. Such an exit will also be called an *unnatural exit* from the body of the procedure, and it does not use the current dump which is used by the natural exit.

A certain collection of information has to be passed to the procedure when a label is an argument. This is called the *value* of the label. This section contains an examination of the nature of this value. Sections 2.5 and 2.6 gave one answer to the question "What does a lambda expression denote?," or in more practical terms, "What collection of information has to be passed to a procedure when it has a procedure as an argument?" The answer given was a *closure*, which, although its exact format may vary, has two parts, a program part and an environment part. This section is an attempt to answer similar questions, namely "What does a label denote?" or "What collection of information must be passed to a procedure when it has a label as an argument?"

There is an analogy between labels and the names of parameterless procedures that deliver no result. A label is similar to a local declaration and is sometimes implemented in the same way as the other declarations in the block. To clarify the analogy between the declaration and use of procedures and labels, a new feature can be added to a programming language to declare the value of a label in the same way as a procedure is declared. This new type of declaration will be introduced by **pp** (short for *program point*) rather than by **procedure.** An example of a block of labeled statements and the equivalent program reexpressed as program point declarations are shown in Fig. 2.20, (a) and (b).

Since a procedure can only be entered naturally, i.e., by using a procedure call, but it is possible to enter a program point by a call, by a **go to** statement, or by falling through from the statement above, the corre-

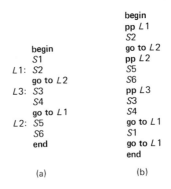

```
                              begin
                              pp L1
                              S2
              begin           go to L2
              S1              pp L2
        L1:   S2              S5
              go to L2        S6
        L3:   S3              pp L3
              S4              S3
              go to L1        S4
        L2:   S5              go to L1
              S6              S1
              end             go to L1
                              end

              (a)             (b)
```

Fig. 2.20 Program point declarations

spondence is not precise. To make it precise, each entry into a program point by continuation from the statement above must be replaced by an explicit **go to** statement. If each **pp** is replaced by **procedure** and all **go to**'s are removed, then the block contains declarations and calls of parameterless procedures. The two actions following the call of a procedure and the call of a program point are identical up to the point of exit. Then the procedure's natural exit is to the place provided for it in the dump that was stored when the procedure was called, but the natural exit from a program point is a natural exit from the block in which it is declared.

A parameterless procedure that produces no results is a very special case of a "parameter*ful*" procedure that *does* produce a result. It is therefore natural to ask if there is a **pp** construction that corresponds to procedures with parameters and delivers results. A new kind of intermediate result which has this property will be introduced. This result, which will be treated as the value of a label, is called a *program closure* and has the following formation rules.

A program closure has a *body* which is a function

and a *dump* which is a state.

A program closure is in many ways analogous to a closure; it will be treated as a function and will be applied to an argument and yield a result. The *body* part of a program closure may be a basic function, a closure, or a program closure. A program closure will be constructed by applying a function called *conspclosure* to a function and a state. The type of a program closure is defined as the type of the function that is its body.

Two additions will be made to the *SECD* machine, 1) to construct and 2) to apply a program closure. When a program closure is applied to its argument the current state, except for the program closure and its argument, is replaced by the state found in the dump of the program closure. The function that is the body of the program closure is then applied to the argument. The state is transformed from

$$(L:x:S, E, apply:C, D)$$

to

$$(body\ L:x:S1, E1, apply:C1, D1)$$
$$\textbf{where}\ S1, E1, C1, D1\ =\ dump\ L.$$

If the body of the program closure is a closure, the next state transformation will store the state that was originally the dump part of the program closure in the current dump. Upon exit from this closure the result will then be passed back to this state, rather than the state at the point of call, in order to resume the computation.

The program closure will be constructed in two stages, so that the state and the function may be chosen separately. A special identifier **J** will be added to the language in order to capture the current dump of the machine. The value of **J** is a function called a *state appender* which contains a state. When a state appender is applied to a function it produces the program closure whose *body* is the function and whose dump is the state found in the state appender. The two state transformations are from

$$(S, E, \mathbf{J}{:}C, D) \qquad \text{to} \qquad (consstateappender\ D{:}S, E, C, D),$$

and

$$(consstateappender\ D1{:}f{:}S, E, apply{:}C, D)$$
$$\text{to}\ (conspclosure(D1, f){:}S, E, C, D).$$

The program closure is most useful for specifying return points in the case of errors. The state is the appropriate return point. The function produces a result which gives the reason for exiting.

It is instructive to compare the actions taken when a closure and a program closure are applied. In the case of a closure, the first action stores the state in the dump. The body is then evaluated and upon exit the result is loaded onto the stack of the reinstalled state. The state transformations are from

$$consclosure(C2, I, E2){:}x{:}S, E, apply{:}C, D$$

to

$$r{:}(),\ (I, x){:}E2, (), (S, E, C, D)$$

where r is the result of applying the closure. The state is then transformed to:

$$r{:}S, E, C, D$$

A shortcut is possible when a program closure that has a closure as its body part is applied. The dump of the program closure becomes the current dump; the control string of the closure becomes the current control; and the argument is added to the environment of the closure and becomes the current environment. If a program closure having the closure above as its function part is applied to the same argument then the state transformation that initiates application is from:

$$\textbf{let}\ L = conspclosure((S1, E1, C1, D1), consclosure(C2, I, E2))$$
$$L{:}x{:}S, E, apply{:}\ C, D$$

to

$$(),\ (I, x){:}E2, C2, (S1, E1, C1, D1).$$

Then the same actions as above are carried out until the exit, when the result r is loaded on to the stack of the state from the program closure, and this state becomes the current state. At the exit the transformation is made from

$$r:(), (I, x):E2, (), (S1, E1, C1, D1)$$

to

$$r:S1, E1, C1, D1.$$

This program closure construction is a genuine addition to the language of expressions, and cannot be defined in terms of application and abstraction. Expressions that contain **J** no longer obey the rules of lambda-conversion. Consider these two expressions which would be convertible if **J** were a pure function.

1. $f(3, 2)$
   where $f(x, y) = (g(x) + y$
       where $g(z) = (sq(L(z, 2))$
          where $L = \mathbf{J}(\lambda(x, y).x^2 - y^2)))$

2. $f(3, 2)$
   where $f(x, y) = ((g(x) + y$
       where $g(z) = sq(L(z, 2)))$
       where $L = \mathbf{J}(\lambda(x, y).x^2 - y^2))$

In the first definition the definition of L qualifies the expression $sq(L(z, 2))$, and in the second case it qualifies the expression $g(x) + y$ **where** $g(z) = sq(L(z, 2))$. Since L only occurs in the first expression, the two are equivalent under lambda conversion rules. However, under the rules for the application of program closures, in the first case the application of L will cause a jump out of the function g, and in the second case it will cause a jump out of the function f. The sq function is not applied in either case. In the first case the stages in the computation are

$$f(3, 2) = g(3) + 2$$
$$= sq(L(3, 2)) + 2$$
$$= L(3, 2) + 2$$
$$= 5 + 2 = 7.$$

In the second case they are

$$f(3, 2) = g(3) + 2$$
$$= sq(L(3, 2)) + 2$$
$$= L(3, 2) = 5.$$

The jumping feature has been added to the language of expressions and makes it possible to terminate the evaluation of an expression abruptly by specifying that the value of the whole expression is to be the value of one of its subexpressions. Consider the expression:

$$(\lambda L. \ldots (L\, x) \ldots (L\, y) \ldots)(\mathbf{J}\, f).$$

Then if L is ever applied to an argument during the evaluation the value of the whole expression is the result of applying f to the same argument. It is possible to reduce the expression above to

$$\ldots (\mathbf{J}\, f\, x) \ldots (\mathbf{J}\, f\, y) \ldots$$

provided that L only occurs at top level in the lambda expression; in this case the expression $(\mathbf{J}\, f)$ has not changed its lambda depth. This second form $(\mathbf{J}\, f\, x)$, in which a program closure is applied immediately after it is constructed, corresponds to a *return* feature found in some programming languages which permits a natural exit from a block or procedure body to be performed in a place other than through the end. In an expression $(\mathbf{J}\, f\, x)$ the program closure is formed from f and the current dump; this is immediately applied giving a result $(f\, x)$ which is then added to the stack of the current dump and reinstalled as the current state. It follows that an exit is made from the procedure that contains $(\mathbf{J}\, f\, x)$ at top level.

This account of jumping is a considerable generalization of the features found in most current programming languages. The fact that the feature has been added to the language of expressions demonstrates that jumping is not necessarily inseparable from statements. By introducing a program closure as the denotation of a label it has also been shown that jumping is not necessarily connected with labeled statements. The label names a program closure in the same way as it names other values. Although the feature may be used to explain the jumping features of programming languages, the extra possibilities provided may also be useful for specifying error exits, failure actions, restart procedures, or backtracking programs. The type of program interrupts that arise as a consequence of the application of an operator in the program may be regarded as a test for failure followed by the application of a program closure. The experiments which have been done suggest that this feature rather than the conventional use of jumping is most useful in these cases.

The following extra possibilities are provided by this general jumping operator.

■ The program closure has a function and a state, whereas the corresponding ALGOL 60 feature has only a state. The corresponding action when a program closure is applied is to apply a function, producing a result which is handed back to a specified state in order to continue the calculation.

■ The state and function may be specified separately, so that it is possible to choose them separately.

■ A new type of value has been introduced which may be embedded in structures. An ALGOL 60 switch, for example, can be regarded as a list of program closures and can be denoted by any expression whose value is a list of program closures.

A program closure may not only be an argument, it may be a result of the application of a procedure and thus permit the possibility of jumping *into* a block in which a label is declared. This possibility is analogous to the situation when a closure occurs as a result of computation. This corresponds to the ability to use a procedure that is declared in an enclosed block.

2.11 SHARING, ASSIGNMENT, AND COPYING

The method for describing sets of data structures introduced in Chapter 1 contains little detail as to representation of the members of the sets inside the computer. The definitions merely create certain functions for constructing and selecting parts of compound information structures together with sets of axioms to which these functions must conform. If it is possible to update a component of a structured object, the representation of structures must be considered in more detail. The *SECD* machine and its extension which adds jumping have been described in terms of these constructing and selecting functions. From this discussion, it follows that it is not possible to update any state component. It has also been assumed that the basic functions are characterized solely by the results of their application to arguments, and that they also do not update any part of the state. A basic function removes its argument from the stack and replaces it by the result; no other state component is changed by a basic function.

An information structure may be represented by a tree of addresses. Each compound structure may be represented in two ways: on the one hand by a segment of contiguous storage cells whose segments contain representatives of its components, and on the other hand by the address of such a segment. A compound information structure can be represented by either the segment or the address of the segment. These have sometimes been called the *L-value* and *R-value,* respectively. The R-value of a structured object might contain L-values of its components. A general rule for representing the data structures which are introduced by a structure definition is to adopt the convention that the structure is represented by an L-value that points to an R-value which contains the fields selected by its named

Fig. 2.21   Representing a list

selecting functions. These will be called its *immediate* or *top level* components. For example, a nonnull list will be represented by an address of a segment that contains an address for the head of the list and an address for the tail of a list, as shown in Fig. 2.21.

The selectors of a data structure can be compounded by functional composition to indicate the positions of components of the structure. Consider the list structure:

$$1, (5, 6), (7, 2).$$

The 5 occupies the position $h \cdot h \cdot t$, and 2 occupies the position $2nd \cdot h \cdot t \cdot t$. Suppose that a function called *update* operates on a position, an object, and a structured object, and that it produces a structure in which the occupant of the position is replaced by the object; for example,

$$update(h \cdot h \cdot t)3(1, (5, 6), 7, 2) = 1, (3, 6), (7, 2)$$
$$update\ 5th\ 10\ (1, 2, 3, 4, 5) = 1, 2, 3, 4, 10.$$

Certain questions about other structures that might be updated are not answered by this definition. These indirect representations give rise to the possibility that two equal components of a structure, occupying distinct positions, might be represented by the same area of memory. In this case, the two positions are said to *share* their components. Clearly if two positions share, so do all their subpositions.

A structure can be pictured as a rooted, directed graph in which each node is an *R*-value, and the edges leading away from a node represent its immediate components. Each of these edges may be labeled with the name of the corresponding selecting function. The set of all paths from the root can be split into equivalence classes under the sharing relation. Two paths from the root (two positions) are equivalent under sharing if both lead from the root to the same node of the graph. The list structure $((a, b), b, c, d)$ can be represented in two ways, as indicated in Fig. 2.22.

In the first graph, the $h \cdot t$ and $h \cdot t \cdot h$ positions share the occupant b. In the second graph, the same two positions, which contain equal occupants, do not share, and there are two copies of b. Updating the $h \cdot t$ or $h \cdot t \cdot h$ positions will have different effects depending on whether the list structure is represented as in (a) or (b). The two results of $update(h \cdot t)e((a, b), b, c, d)$ are (a) $(a, e), e, c, d$ and (b) $(a, b), e, c, d$, respectively.

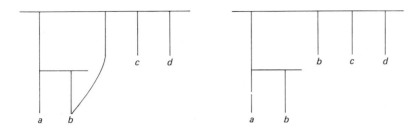

Fig. 2.22   Shared and unshared list structures

In the first case both the $h \cdot t$ and $h \cdot t \cdot h$ positions are updated simultaneously, in the second case only the position $h \cdot t \cdot h$ is updated. To be more precise we have shown the results as graphs in Fig. 2.23, in which the sharing relation as well as the occupant of each position is specified.

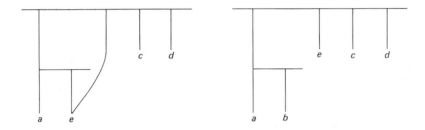

Fig. 2.23   Graphs of sharing relations

To precisely specify the change to the machine state that contains updating operations, we must specify the change to the sharing relation as well as the change to the occupant of each position of the state. A rule for indicating the change to the sharing pattern will be established that will be applicable to a function which operates on and produces list structures.

Consider the following list structure transforming function:

$$f(a, b, (c, d), e) = (e, g, (d, e, a)).$$

A list of pairs of positions can be obtained from this definition. The first member of each pair is the position of an identifier in the bound variable; the second is the position of the same identifier in the body or definiens of the function. The identifier a occurs in the position *first* in the bound variable and in the position *third • third* in the body. The complete list of pairs follows; each pair is labeled with the corresponding identifier.

(*a*) *first, third • third*

(*b*)

(*c*)

(*d*) *second • third, first • third*

(*e*) *fourth, first*

(*e*) *fourth, second • third*

The occupants of the positions corresponding to *b* and *c* do not survive the application of the function. The identifiers in the bound variable should be distinct. Such a pairing of before and after positions will specify a change to the sharing pattern according to the following rule. If

$$(p_1, q_1), (p_2, q_2), \ldots (p_n, q_n)$$

is a pairing of this sort, and if the positions p_i, p_j share in the argument of the function; then q_i and q_j share in the resulting list structure. This is also true of the subpositions. More generally, if $r \cdot p_i$ and $s \cdot p_j$ share in the argument; then the positions $r \cdot q_i$ and $s \cdot q_j$ share in the result. The convention adopted thus permits list transformations to be done with a minimum of copying; only *L*-values not *R*-values are copied.

Consider the application of the function

$$f(a, b, (c, d), e) = e, g, (d, e, a))$$

to the list structure $(A, X), B, (C, (X, Y)), (Y, D, E)$ whose sharing relation

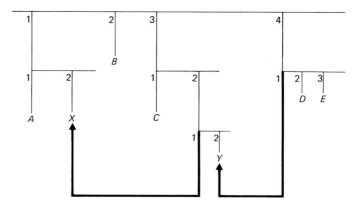

Fig. 2.24   The graph of the argument

is represented by the graph in Fig. 2.24. The following positions share in the argument list structure:

$$(X)\ second \cdot first \text{ shares with } first \cdot second \cdot third$$
$$(Y)\ second \cdot second \cdot third \text{ shares with } first \cdot fourth$$

The resulting list structure may be written as follows:

$$((Y, D, E), f, ((X, Y), (Y, D, E), (A, X)))$$

The next consideration is the sharing pattern in this result. In the argument list structure, *second · first* shares with *first · second · third*. In the function, *first* is paired with *third · third*, and *second · third* is paired with *first · third*. It follows that positions *second · third · third* and *first · first · third* share in the result. By a similar argument, since any position shares with itself the two positions containing *e*(*first* and *second · third*), positions *first · first* and *second · first · third* share in the result. The resulting graph is shown in Fig. 2.25.

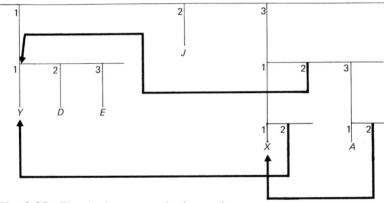

Fig. 2.25   The sharing pattern in the result

The state of the *SECD* machine is now treated as a list structure in order to specify the effect of assignment. The assignment operator is applied to a position in the state (to be more precise, the environment) and an object, and changes not only the occupant of the position but also the occupants of all positions that share with the updated position. This is conveniently performed by operating on the *L*-value and updating the corresponding *R*-value; all other equal *L*-values in the state become simultaneously updated.

     The assignment statement, written $a := b$, requires the expression a to have a value that is a position in the state. It also requires that the

R-values of *a* and *b* have the same shape. The assignment statement causes the *R*-value of *a* to be overwritten by the *R*-value of *b*. Assignment requires an additional transition rule to be added to the *SECD* machine. The new operator : = is applied to a pair of *L*-values. When '*apply*' is at the head of the control string and : = is at the head of the stack, the following transformation takes place:

from (*S, E, C, D*)

to (*update*(*first* • *second* • *Ss*)(*second*(*second S*))(*S, E, C, D*).

In this, the stack, environment, control and dump positions of the state are written *Ss, Es, Cs,* and *Ds,* respectively. This is then followed by an instruction which loads the value of *b* to the stack, namely:

$$\lambda((a, b):S, E, C, D).b:S, E, C, D.$$

The sharing pattern is also changed by functions that copy their arguments. Suppose that there is a function called *separate* which operates on an *L*-value and produces a copy of the *R*-value pointed to as its result. Iteration of this function can copy all the segments of a structure. In terms of the equivalence relation, the function *separate* removes a position from an equivalence class, and produces a new class containing one member.

The consequences of particular choices of the change to the sharing pattern on the effect of assignment operators in the *SECD* machine will be examined next.

1. When the expression is an identifier *Y*, the state changes from (*S, E, Y:C, D*) to (*location E Y E:S, E, C, D*), which means that the value of an identifier is an environment position, or *L*-value. In the resulting state the *h.Ss* position is left sharing with the (*location E Y*) • *Es* position of the state. Thus the value of any identifier, bound or free, is a possible argument for the left side of : =. In other words, the choice has been made to pass arguments in the same way as they are passed to FORTRAN subroutines. This method is sometimes known as a "call by reference."

2. If the head of the control is a lambda expression the step is from

(*S, E, conslambda*(*x, M*):*C, D*) *to* (*consclosure*(*M, x, E*):*S, E, C, D*)

and the current environment shares with the environment of the closure in the resulting state. In other words the positions *Es* and *Ec* • *h* • *Ss* share in the resulting state. The environment, bound variable, and control positions of a closure will be written *Ec,Bc* and *Cc*, respectively.

3. If *apply* is found at the head of the control string, and a closure is found at the head of the stack; the step from

(*consclosure*(*C*1*, x, E*1):*y*:*S, E, apply*:*C, D*)

to ((), (*x, y*):*E*1, *u C*1, (*S, E, C, D*))

is taken. In the resulting state, the tail of the environment $(t \cdot Es)$ shares with the environment of the closure, and any of its cosharers. This, together with (2.) above, ensures that the environment of a closure, and hence the environment that was current when it was constructed, may be updated via a free variable of the body when the function is applied. Such an updating is usually a *side effect* of the application of the function or procedure. The new $second \cdot h \cdot Es$ shares with all cosharers of the old $second \cdot Ss$ and so the application of a function may also have a side effect through its argument (y).

4. The step taken upon exit from a closure is from $(S, E, C, (S1, E1, C1, D1))$ to $(h\,S : S1, E1, C1, D1)$. The $h \cdot Ss$ of the resulting state shares with the $h \cdot Ss$ of the old state, and the $t \cdot Ss, Es, Cs,$ and Ds of the new state share with the $Ss \cdot Ds, Es \cdot Ds, Cs \cdot Ds,$ and $Ds \cdot Ds$ positions of the old state. If the result $(h\,S)$ is a closure constructed in the environment E, then E survives indirectly by being part of the closure. This feature is banned from most programming languages. If it is banned, environments and dumps can be stored and discarded in a last-in-first-out manner, in the same way as is the stack.

5. The two actions taken in constructing a program closure are first to capture a state and then to attach it to a function. Therefore the dump in the state appender shares with the dump current at the time of its construction. Any change to the environment of this dump will indirectly change the dump in the state appender. The dump in the state appender also shares with the dump in the program closure which is constructed when it is applied. Again, any change to the environment part of the dump of the program closure will change the program closure indirectly.

6. The change to the sharing pattern made by each basic function must be specified.

7. The value of every expression is an L-value and is suitable as the left-hand side of an assignment statement. Although statements like

$$a + b := \ldots$$

are possible, they are only useful if the value of the left-hand side shares with some other position in the state.

It should be noted that after a closure has been constructed it can be changed by an assignment to the part of its environment that it shares with other closures. This interpretation is usually assumed in ALGOL 60, although it is never explicitly stated. Some alternative strategies are possible. One might adopt the view, for example, that once a closure has been constructed it cannot be changed. In order to implement this strategy the environment of the closure must be guarded against assignment statements by copying it at the time of construction. Once this is done, the environment can only be changed by assignment statements in the control string of the closure, and such a closure cannot assign to the environment of any other closure. Another strategy is to make a copy of the closure at the time of its application. In this case, the closure environment can be changed between the time of its construction and the time of its first application or even between two successive applications. The application of a closure whose environment is changed at the time of application cannot change the environment of another closure. If the closure is copied both at the time of its construction and at the time of each application then its environment can only be changed by its own control string.

The standard method of passing arguments by reference is used in FORTRAN, but in ALGOL 60 two other strategies, *call by value* and *call by name,* are followed. In the call-by-value discipline of passing arguments to procedures, the argument must first be evaluated, and a *copy* of its value then passed to the procedure. In this case an assignment through the variable causes the copy to be updated. In the call-by-name discipline, the argument expression is evaluated at each point of use within the procedure. An argument that is called by name can be implemented in the *SECD* model by changing each argument called by name into a parameterless procedure. Each occurrence of the variable in the body is changed to the application of the variable to a null argument list.

2.12   OTHER METHODS OF EVALUATING EXPRESSIONS

All the machines described so far adopt the strategy of applying functions to evaluated arguments. This section contains a number of alternative strategies in which certain calculations are postponed or advanced.

Delayed evaluation.   The device of changing operand expressions into lambda expressions was used to delay the evaluation of the arms of conditional expressions and to explain the call-by-name feature of ALGOL 60. The new strategy will be to postpone the evaluation of a subexpression until its value is actually needed. If it is not needed then its evaluation is completely evaded.

There can be two types of basic functions in the system; a basic function either: 1) requires that its arguments be evaluated before it is applied, e.g., +, or 2) may be applied to both evaluated and unevaluated arguments, e.g., *prefix*. The system must know whether each function tolerates unevaluated arguments or not. It must also know for each object whether the argument is evaluated or not. The arguments to each closure can be left unevaluated. It is up to the functions within the closure to initiate evaluation if necessary.

At first sight the delayed evaluation strategy for arguments might seem to be inefficient because a delayed argument has to be evaluated each time its argument is needed. If pure expressions, (without **J** or *assignment*) are evaluated, then this value must be the same each time. In this case it is possible to arrange that the expression is only evaluated once, and that second and subsequent uses refer to this value. If the expression is to stand in for its value then it must carry with it both the expression itself and the environment which gives values to its free identifiers. When an operator/operand combination is encountered the operand must be paired with the current environment producing what is called a completed closure, marked unevaluated. The operator is evaluated and applied to this completed closure. If the function requires an evaluated argument, then the completed closure is evaluated in a way similar to the application of a closure to its argument. The current state is stored in the dump and the new E and C components of the state come from the completed closure. There is a slight difference in that when the value is obtained it must be used to update the completed closure, so that all references to the completed closure are replaced by references to its value. This means that all completed closures and all objects in the system must be represented indirectly by an L-value whose R-value is capable of holding a completed closure and is unevaluated or can hold any object and is evaluated. The normal assignment operator will then change all objects containing references to this R-value.

This method of delaying the evaluation of operand expressions can be summarized by changes to the $SECD$ machine in the combination arm, the application of a basic function, and the exit arm. When a combination is encountered at the head of the control, a completed closure is loaded to the stack composed from the operand and the current environment. When an intolerant basic function is applied to a completed closure it must initiate its evaluation and reapply the function to the result. The position that has to be updated will be stored in the dump. A new type of dump, an assigning dump, which contains a state and a position, must be introduced. When the control string is exhausted and the assigning dump is in the current dump position, the system acts to both install the state part of the dump as the current state and to assign the result to the position found in the dump. All positions of the completed closure in the state are simultaneously updated. The revised $SECD$ machine transition for the *procrastinating machine* follows.

> *transition*$(S, E, C, D) =$
>> **if** *null C*
>>> **then if** *assigning D*
>>>> **then** $(y: = r)(S1, E1, C1, D1)$

$$\textbf{where } y, (S1, E1, C1, D1) = D$$
$$\textbf{and } r = h\ S$$
$$\textbf{else } (h\ S{:}S1, E1, C1, D1)$$
$$\textbf{where } S1, E1, C1, D1 = D$$
$$\textbf{else let } X = h\ C$$
$$\textbf{if } identifier\ X$$
$$\textbf{then } (location\ E\ X\ E{:}S, E, t\ C, D)$$
$$\textbf{else if } lambda\ expression\ X$$
$$\textbf{then } consclosure(body\ X, bv\ X, E){:}S, E, t\ C, D$$
$$\textbf{else if } X = \text{`}apply\text{'}$$
$$\textbf{then let } f{:}y{:}S1 = S$$
$$\textbf{if } closure\ f$$
$$\textbf{then let } consclosure(C1, J, E1) = f$$
$$(), (J, y){:}E1, u\ C1, (S1, E, t\ C, D)$$
$$\textbf{else if } sensitive\ f \wedge uneval\ y$$
$$\textbf{then } (), E1, C1, (y, (S, E, C, D))$$
$$\textbf{where } consuneval\ E1\ C1 = y$$
$$\textbf{else } f\ y{:}S1, E, t\ C, D$$
$$\textbf{else let } combine(F, A) = X$$
$$(consuneval\ E\ (u\ A)){:}S, E, F{:}\text{`}apply\text{'}{:}t\ C, D$$

Although the result of evaluating a completed closure may still be un-evaluated, it is presented again to the same function and is repeatedly evaluated until its value appears.

Although the order of evaluation of subexpressions is different, the procrastinating machine produces the same value as normal evaluation when applied to expressions of the λ-**I** calculus. Some expressions of the λ-**K** calculus, which have no value under normal evaluation because the evaluation of an operand is nonterminating, will be given a value by the procrastinating machine if the value of the operand is not required. The procrastinating machine may be valuable for testing an incomplete program. The expressions may contain subexpressions which have no value. The expression will be evaluated provided its value does not depend on the values of these subexpressions.

The strategy discussed here has been to delay evaluation as long as possible. Other strategies which are a compromise between this and normal evaluation are possible. It is possible to arrange that the evaluation of a subexpression be initiated earlier than needed. This has the advantage of freeing some applications of inner functions from the burden of recognizing unevaluated arguments.

Parallel evaluation.   The evaluation of the operator and operand could be initiated simultaneously and proceed concurrently, using two machines. This involves the same amount of computation as in normal evaluation, but the computation may be spread over several machines so as to reduce the elapsed evaluation time. In this case each object is either evaluated or in the process of being evaluated; there is no possibility of an unevaluated object. The actions may be associated with the apply operation. The apply operator must wait until the operator is evaluated and then two strategies are possible. Either the program could wait until the operand is evaluated before applying the operator, or it could apply the operator to an unevaluated argument delegating the decisions to the operator. In the more cautious strategy, a tree of activities is constructed which are all held up except those at the ends of the tree. When two subtrees are evaluated the tree is pruned and activity can begin at their common root. In the less cautious or *rash* machine it may be possible to have a dead activity with a living offspring; but since more than one activity might be held up waiting for the same task to terminate, it is not valid to kill an offspring when its parent is terminated. The result may still be required elsewhere.

REFERENCES AND BIBLIOGRAPHY

This chapter is based upon Landin's SECD machine, an interpreter for expressions in Church's lambda notation. This, together with the extensions for jumping and assignment, is both a model for a family of programming languages and one method of explaining some semantic features of other programming languages. The entire chapter is summarized in Landin's short note [2–12]. Other strong influences are McCarthy's LISP, and Dijkstra's implementation [2–7] of ALGOL 60. The use of pushdown lists has a long history, and several computers have incorporated them into the hardware. Other programming languages based to a greater or lesser extent on the lambda calculus notation have been developed. It is surprising how much of ALGOL 60 can be explained in terms of the lambda-calculus notation [2–11], since ALGOL 60 was not designed from that point of view. It seems that this is one instance where the theoretical studies of computation and the design of a system for practical programming purposes have arrived at the same place. The possible uses of function-producing functions have not been explored to any great extent to date for practical purposes. They promise to be useful because they make it possible not only to write programs but to write programs that combine other programs. Some uses will be explored in Chapter 3.

2–1. Allmark, R. H., and J. R. Lucking, "Design of an arithmetic unit incorporating a nesting store," (in) *Proc. IFIP Congress 62,* Amsterdam: North Holland, 1963, pp. 694–698.

2–2. Barton, R. S., "A new approach to the functional design of a digital computer," *Proc. WJCC,* ACM New York, 1961, pp. 393–396.

2–3. Bauer, F. G., "The formula controlled logical computer 'Stanislaus,' " *Math. Comp.,* Vol. 14, No. 69, 1960, pp. 64–67.

2–4. Burge, W. H., "The evaluation, classification and interpretation of expressions," *Proc. 19th National ACM Conf.,* New York, 1964, pp. A1.4.1-1–1.4.22.

2–5. Burge, W. H., "Interpretation, stacks and evaluation," (in) *Introduction to Systems Programming,* (ed.) P. Wegner, London and New York: Academic Press, 1964, pp. 294–312.

2–6. Burks, A. W., D. W. Warren, and J. B. Wright, "An analysis of a logical machine using parenthesis-free notation," *M.T.A.C.,* Vol. 8, No. 46, 1954, p. 53.

2–7. Dijkstra, E. W., "Recursive programming," *Num. Math.,* Vol. 2, No. 5, 1960, pp. 312–318.

2–8. Hamblin, C. L., "Computer languages," *Austr. J. Sci.,* 1957, pp. 135–139.

2–9. Hamblin, C. L., "Translation to and from Polish notation," *Computer J.,* Vol. 5, No. 3, 1962, pp. 210–213.

2–10. Landin, P. J., "The mechanical evaluation of expressions," *Computer J.,* Vol. 6, No. 4, 1964, pp. 308–320.

2–11. Landin, P. J., "A correspondence between ALGOL 60 and Church's lambda-notation," *CACM,* Part 1: Vol. 8, No. 2, 1965, pp. 89–101; Part 2: Vol. 8, No. 3, 1965, pp. 158–165.

2–12. Landin, P. J., "An abstract machine for designers of computing languages," *Proc. IFIP Congress 65,* Vol. 2, Washington: Spartan Books, London: MacMillan and Co., 1966, pp. 438–439.

2–13. Myamlin, A. N. and V. K. Smirnov, "Computer with stack memory," *Proc. IFIP Congress 68,* Amsterdam: North Holland, 1968.

2–14. Samelson, K., and F. L. Bauer, "Sequential formula translation," *CACM,* Vol. 3, 1960, pp. 76–83.

2–15. Randell, B. and L. J. Russell, *ALGOL 60 Implementation.* London and New York: Academic Press, 1964.

3
Data
Structures

3.1 INTRODUCTION

In the early stages of writing a program it is useful to have a systematic method of specifying both how the program is to be divided into its parts and how these parts are interrelated. Before any part is programmed in detail, it is important to decide the purpose of each subroutine, and to determine whether that subroutine is being used in a proper context. A written record of this overall view of the entire program throughout the program-writing stages is a valuable record, since when changes are necessary, it permits the effects of each change to be clearly seen.

A subroutine is usually limited to accepting arguments that belong to a particular set called the *domain* of the subroutine and producing members of a particular set called its *range of values* or *range*. A first step in the orderly development of a large program is to determine the domain and range of the entire program and of each of its subroutines. A method for describing tree-like data structures was given in Chapter 1. This chapter contains a systematic way to obtain programs which match the data structure.

We have attempted here to separate decisions about the logical structure of the information from decisions about how the structures are to be physically represented. This separation is valuable, because the shape of the structures needed depends on the problem being solved; whereas the choice of physical format depends more on the operations and data structures provided by the computer or programming system being used. In practice, the distinction between logical and physical structure is difficult to make; and decisions about logical structure are usually made by default. The physical representation is often chosen at too early a stage and wholesale changes of representation become impossible without rewriting large parts of the program. The effort of retaining the distinction seems worthwhile, however, because a clear picture of a set of information structures implies to some extent the shape of the programs that operate on, or create members of, the set.

In some cases the underlying data structure may be hidden and take the form of arithmetical calculations performed by the instructions in the program. It seems preferable, especially in the early stages, to make this structure explicit so that the programs that operate on the structure are easy to understand.

A technique for constructing programs that match the data structure is first to write a mere skeleton of a program and later to flesh it out by supplying arguments that specify the actions to be taken within the skeleton. The skeletal program specifies the whole family of programs obtained in this way and any properties of the general-purpose program are then applicable to all the members of this family.

All the programs chosen to illustrate the technique in this chapter operate on rooted, ordered trees. The general purpose skeleton embodies the method of scanning the structure, and the arguments supplied specify the actions to be taken during the scanning. The general purpose programs can be produced in a mechanical fashion from the description of the set of trees (sometimes called the *abstract syntax* of the set) being scanned. The set of trees considered have components that are either *atomic,* which means that their internal structure is not examined by the general-purpose program, or *composite,* and their internal structure is so examined.

A set of data structures can be specified without being committed to the nature of its atomic components. It is possible, for example, to define the set of lists and to write functions on lists without specifying the nature of the list elements, which will be treated as atomic as far as the general-purpose list function is concerned. The function that is concerned with the nature of the list elements will be supplied as an argument to the list function. A list element can itself have a structure, for example, it can be some sort of tree containing atomic components of its own. It is therefore possible to create a general-purpose function for a list of trees from the corresponding functions for lists and trees by supplying one as an argument to the other.

A method is introduced for representing data structures by *streams,* which are similar to coroutines in producing the components of the structure only when they are required. It is often easier to understand and program a sequence of passes which happen one after the other than to consider an involved process in which different things are mixed together in a program. On the other hand, the mixture is often the more efficient program. It is possible to get the best of both worlds in the sense that the program is written as if it were a multipass program but, when executed, the parts of the separate passes are interleaved. Similarly, it is often easier to consider the sequence of values taken on by a variable in a program as an object of computation rather than to consider the mechanisms that use, test, and change the value of that variable. The coroutine technique enables the pieces of program which contain the variable to be gathered together in one place, rather than being scattered about the program. This stream, or coroutine, technique is used to create sets of combinatorial configurations. These sets are most easily specified in a tree-like manner but the program produces the members one at a time because a stream of configurations is produced.

3.2 DATA STRUCTURES

In chapter one, a set of structures was introduced both by naming its components and by specifying the type of each component. If there are alternative formats for the same set, then a predicate is associated with each set and given a name. The definition of an expression having infixed operators, for example, can be written:

- An expression

   is either *atomic* and is an identifier

   or *compound*

   and has an *operator,*

   which is an infixed operator,

   and a *right* and a *left,*

   both of which are expressions.

In this definition, the names of the predicates and selectors are italicized. Constructing operators are also introduced. For example, to construct a compound expression from two expressions and an infixed operator, an operator must be introduced. The English sentences are so stylized that they constitute a notation in themselves. It is possible to present the same information in an alternative notation which, although perhaps harder to read, has some benefits because it is more precise and permits almost immediate acquisition of descriptions of objects that are closely related to the set of structures.

Abstract syntax. It is possible to separate the definition of the set itself from the naming of its associated predicates, selectors, and constructors. For this purpose, two operations on sets, *Cartesian product* (written *cp*), and *disjoint union* (written *du*) are introduced.

The Cartesian product of the sets *A, B, C,* and *D* is written

$$cp(A, B, C, D)$$

and is the set of all 4-tuples whose first member is of type *A,* second is type *B,* third is type *C,* and fourth is type *D.* The notation is an alternative to the usual form

$$A \times B \times C \times D$$

and is a specification of the set of structures having four components, whose types are specified but whose selecting functions are unnamed at this point. The closest approximation to its English language equivalent is:

> has a ... which is an *A*
>
> and a ... which is a *B*
>
> and a ... which is a *C*
>
> and a ... which is a *D*

The prefixed *cp* notation is used because it permits one-component and zero-component structures to be defined. These will be written *cp*(*u A*) and *cp*(), respectively. The component selectors can be obtained and named as follows. Suppose that the programming system accepts definitions of sets such as

$$\textbf{def } complex = cp(real, real)$$

and stores the definition of this set in a way that permits functions to be applied to it. A function called *selectors* is used which when applied produces a list of the selectors. At a later stage names can be given to the selectors by the definition:

$$\textbf{def } real, imaginary = selectors\ complex.$$

A function that generates constructors, called *construct,* is also needed. It is applied to a type and a list of objects and produces one object of the correct type having the listed objects as its components. Using *construct* with the expression

$$construct\ complex\ (5.3, 7.2)$$

a given complex number can be constructed. Alternatively, a complex number constructing function can be defined

$$\textbf{def } conscomplex = construct(complex)$$

which can be applied directly to a pair of real numbers to produce a complex number (i.e., *conscomplex*(5.3, 7.2)). The selecting and constructing functions are closely related. If the two components of a complex number are selected, and a complex number is then constructed from them, the result must be equal to the original. In other words

$$x = construct\ complex(real\ x, imaginary\ x).$$

In general, the axioms which express the fact that a structured object can be constructed in only one way, take the following form. If C_1, C_2, C_3, ..., C_n are types, and

$$\textbf{def}\ S\ =\ cp(C_1, C_2, C_3, \ldots, C_n)$$
$$\textbf{def}\ S_1, S_2, S_3, \ldots, S_n = selectors\ S$$

then

$$S_k\ (construct\ S\ (X_1, X_2, X_3, \ldots, X_n)) = X_k$$

and

$$X = construct\ S\ (S_1\ X, S_2\ X, S_3\ X, \ldots, S_n\ X),$$

provided X is of type S.

The disjoint union of two sets A and B, written $du(A, B)$, contains all the members of A and B, but each item is assumed to be labeled with the name of the set from which it came. It is therefore possible to determine whether a member of $du(A, B)$ came from A or B. The English language equivalent of:

$$du(A, B, C, D)$$

is

- is either ... and is A

   or ... and is B

   or ... and is C

   or ... and is D.

A function called *predicates* will be used to produce the predicates for a set that is a disjoint union. The function *du* will be extended so that it operates on a list of sets, rather than on a pair. Using this notation, the definition of an expression having infixed operators becomes

$$\textbf{def rec}\ expression\ =$$
$$du(identifier,$$
$$cp(expression, infixed\ operator, expression))$$

and the predicates for an expression can now be defined as

$$\textbf{def}\ atomic, compound = predicates\ expression.$$

The arguments of *du* are ordered, and the resulting list of predicates are in the same order. The first predicate tests whether an item is the first alternative, the second tests for the second alternative, etc. It is convenient at this stage to make use of yet another function, called *parts* which is an inverse to the disjoint union operator.

$$parts(du(C_1, C_2, \ldots, C_n)) = C_1, C_2, \ldots, C_n$$

Now it is possible to introduce the sets of *atomic* and *nonatomic* expressions, as follows:

def *atomic, nonatomic = parts expression*

and the selectors for *nonatomic* expressions:

def *left, operator, right = selectors nonatomic*

The axioms that express the fact that the predicates are effective are

$$T = du(C_1, C_2, \ldots, C_n).$$

If

$$P_1, P_2, P_3, \ldots, P_n = predicates\ T$$

and

then
$$P_i X = \begin{cases} \textbf{true } \textit{if } X \varepsilon C_i \\ \\ \textbf{false } \textit{otherwise.} \end{cases}$$

Functions.   There are very close analogies between the methods used for structuring data and those used for structuring the function which is applicable to that data. If A is a set, then its members can occur as x in the context $F_A(x)$, where F_A is a function that is applicable to an A. The set of all functions that are applicable to a member of $cp(A, B, C)$ whose selectors are *first, second,* and *third* may be expressed as follows:

$$\lambda x.F(F_A\ (first\ x), F_B\ (second\ x), F_C\ (third\ x))$$

in which the functions F_A, F_B, and F_C are applied to the components and the results are combined by using a function F. The set of functions that are applicable to members of $du(A, B, C)$, whose predicates are *is-A, is-B,* and *is-C,* may be expressed as

$$\lambda x. \textbf{if } is\text{-}A\ (x)$$
$$\textbf{then } F_A\ (x)$$
$$\textbf{else if } is\text{-}B(x)$$
$$\textbf{then } F_B\ (x)$$
$$\textbf{else } F_C\ (x)$$

in which tests are first made for the alternative formats and a function of the correct type is then applied.

If the definition of the structure is self-referential, e.g.:

$$\textbf{def rec } A = \ldots A \ldots A \ldots$$

Then the simplest function on the members is also self-referential:

$$\textbf{def rec } F_A = \ldots F_A \ldots F_A \ldots .$$

The same function is applied to each component of type A as is applied to the whole structure.

Enumerating generating functions.   If $n(A)$ is the number of members of A; then $n(cp(A, B)) = n(A) \times n(B)$, $n(cp(A)) = n(A)$, and $n(cp()) = 1$. The number of members of $du(A, B)$, i.e., $n(du(A, B))$, is equal to $n(A) + n(B)$. It is possible to provide more information about these sets as follows. Each object under consideration either has components or is atomic and has no internal structure. The members of each set can be classified by the number of atomic components they contain. This classification can be expressed as an *enumerating generating function* in which the coefficient of x^n is the number of members of the set having n atomic components. Suppose that $A(x)$ and $B(x)$ are two generating functions of this sort for the two sets A and B, respectively,

$$A(x) = \Sigma a_n x^n, B(x) = \Sigma b_n x^n$$

in which a_n is the number of members of A having exactly n atomic components of a certain type, and b_n is the number of members of B having n components of the same type. Then the number of members of $cp(A, B)$ having exactly n atomic components is the coefficient of x^n in

$$A(x)B(x) = \Sigma \Sigma a_k b_{n-k} x^n$$

In this sum, the term $a_k b_{n-k}$ is the number of members of $cp(A, B)$ in which the first contributes k atomic components and the second contributes $n - k$. The number of members of $du(A, B)$ having n atomic components is $a_n + b_n$. The generating function for $du(A, B)$ is therefore found by adding the generating functions for A and B:

$$A(x) + B(x) = \Sigma(a_n + b_n)x^n.$$

The same rules for constructing generating functions are applicable to sets of trees having more than one type of atomic component and multivariable generating functions. The generating function of a Cartesian product is the product of the generating functions of the component sets, and the generating function of a direct union is the sum of the generating functions for the alternative sets. The analogies between sets of structures, functions, and generating functions are summarized in the table below and will be explored by a number of examples in the following sections.

Sets	Program	Generating function
$cp(A, B)$	$F(F_A\,(first(x)),$ $F_B\,(second(x))))$	$A(x) \cdot B(x)$
$du(A, B)$	**if** is-A **then** $F_A(x)$ **else** $F_B(x)$	$A(x) + B(x)$
$A = \ldots B \ldots B \ldots$	$F_A = \ldots F_B \ldots F_B \ldots$	$A(B(x))$

In the last line, $A = \ldots B \ldots B \ldots$ means a set A is defined in terms of a set B; $F_A = \ldots F_B \ldots F_B \ldots$ means that the function F_A is defined in terms of the function F_B; and $A\,(B(x))$ means that $B\,(x)$ is substituted for x in $A\,(x)$.

3.3 SOME SIMPLE STRUCTURES

1. At first sight the structure having no components might not seem to be very useful. However, two or more can be made into arguments of a disjoint union to form a switch. The description of the set is $cp()$. Its generating function is x^0 or 1, signifying that it has no components. A two-way switch or truth value can be defined as follows.

$$\textbf{def } truthvalue = du(cp\,(), cp\,())$$
$$\textbf{def } truth, falsehood = parts\ truthvalue$$
$$\textbf{def true} = construct\ truth\,()$$
$$\textbf{def false} = construct\ falsehood\,()$$
$$\textbf{def } istrue, isfalse = predicates\ truthvalue$$

One of the contexts in which an expression describing a truth value occurs most frequently is in the condition part of a conditional expression. If x is the variable for **true** and y is the variable for **false**, then the generating function for truth values is $x + y$. If $x + y$ is substituted for a variable in another generating function then the coefficient of x^n will count the number of structures containing n **true**'s; and that of y^n, the number of structures containing n **false**'s. The most frequently used operations on truth values are *or, and,* and *not.* Their definitions are given in the examples in Chapter 1.

2. In a similar way, a set having five alternatives can be defined as follows:

$$du(cp(), cp(), cp(), cp(), cp()).$$

If it is possible to map the alternatives into the integers 1 to 5, an expression could be used in the context of a case expression or case statement, such as:

$$\textbf{case } X \textbf{ of } S\,1;$$
$$S\,2;$$
$$S\,3;$$
$$S\,4;$$
$$S\,5$$

The only functions applicable to members of sets of this kind are: a) those which construct one of the alternatives and b) their predicates.

3. The generating function for $du(A, B, C, D, E)$ may be written $x_1 + x_2 + x_3 + x_4 + x_5$, in which the variables x_1, x_2, x_3, x_4, x_5 are used for items of types A, B, C, D, E, respectively.

4. Similarly, the generating function for $cp(A, B, C, D, E)$ is $x_1 \times x_2 \times x_3 \times x_4 \times x_5$.

5. A function called *possible* can be defined as follows and can be used to describe a field of a record which may or may not be present.

$$\textbf{def } possible\,A \ = \ du(A, cp())$$
$$\textbf{def } exists, notexists \ = \ predicates(possible\,A)$$

The set (*possible A*) is used in the context

$$\textbf{if } exists \ x \ \textbf{then } f(x) \ \textbf{else } a$$

in which only the first arm can contain a reference to x. The associated generating function is

$$possible(x) \ = \ 1 + x.$$

6. The expression $cp(u\,A)$ denotes the set of structures having one component of type A. The only operations permitted are to construct one structure or to select its sole component.

$$\textbf{def } ref\,A \ = \ cp(u\,A)$$
$$\textbf{def } addr(x) \ = \ construct\,(ref\,A)\,x$$
$$\textbf{def } content \ = \ selectors\,(ref\,A)$$

The *addr* function produces one level of indirect reference. The *contents* functions produces an object of type *A* from an object of type $cp(u\ A)$.

7. The set of integers can be defined as follows:

$$\textbf{def } integer\ =\ du(cp(),\ cp(u\ integer))$$
$$\textbf{def } zero,\ nonzero\ =\ predicates\ integer$$
$$\textbf{def } zerointeger,\ nonzerointeger\ =\ parts\ integer$$
$$\textbf{def } successor\ x\ =\ construct\ nonzerointeger(u\ x)$$
$$\textbf{def } 0\ =\ construct\ zerointeger\ ()$$
$$\textbf{def } predecessor\ =\ first(selectors\ nonzerointeger)$$

If *x* is an *integer* and *y* is a *nonzero integer*

$$zero\ 0\ =\ \textbf{true}$$
$$zero(successor\ x)\ =\ \textbf{false}$$
$$nonzero\ 0\ =\ \textbf{false}$$
$$nonzero(successor\ x)\ =\ \textbf{true}$$
$$successor(predecessor\ y)\ =\ y$$
$$predecessor(successor\ x)\ =\ x$$

The family of functions most naturally associated with this definition has already been encountered in Chapter 1, namely:

def rec *R a g n* =
> **if** *zero n*
> **then** *a*
> **else** *g n*(***R*** *a g* (*predecessor n*))

8. A data structure having a structure similar to that of the integers is the set of indirect references.

def rec *indref A* = $du(A,\ cp(u(indref\ A)))$.

The family of functions that correspond to an indirect reference is, using the same predicates and selectors as for integers,

def rec *R f g x* =
> **if** *zero x*
> **then** *f x*
> **else** *g* (***R*** *f g* (*predecessor x*))

3.4 LISTS

A list is the most commonly used data structure in programming. The definition of a list of elements of type *A* is:

- An *A-list*
     is either *null*
     or has a *head* which is an *A*
          and a *tail* which is an *A-list*

or

$$\textbf{def rec } list\ A\ =\ du(cp(),\ cp(A,\ list\ A\,))$$

and the operations on lists can be defined as follows:

$$\textbf{def } null,\ nonnull\ =\ predicates\ (list\ A\,)$$
$$\textbf{def } nullc,\ nonnullc\ =\ parts(list\ A\,)$$
$$\textbf{def } h,\ t\ =\ selectors\ nonnullc$$
$$\textbf{def } prefix\ x\ y\ =\ construct\ nonnullc(x,\ y)$$
$$\textbf{def } nullist\ =\ construct\ nullc\ ()$$

in which *head* has been abbreviated to *h*, and *tail* to *t*. The expression $x:y$ is used as an abbreviation for *prefix x y*, () to denote the null list, *a, b, c, d, e* instead of $a:(b:(c:(d:(e:()))))$, and *u x* to denote a list with one element. The types of the list functions introduced are:

$$null\ \varepsilon\ A\text{-}list\ \rightarrow\ truth\ value$$
$$head\ \varepsilon\ nonnull\ A\text{-}list\ \rightarrow\ A$$
$$tail\ \varepsilon\ nonnull\ A\text{-}list\ \rightarrow\ A\text{-}list$$
$$prefix\ \varepsilon\ (A\ \rightarrow\ (A\text{-}list\ \rightarrow\ A\text{-}list))$$
$$()\ \varepsilon\ A\text{-}list$$

These functions are closely interrelated, as for instance,

$$null()\ =\ true$$
$$null(x:y)\ =\ false$$
$$h(x:y)\ =\ x$$
$$t(x:y)\ =\ y$$
$$(h\ z):(t\ z)\ =\ z$$

in which *x* is an *A*, *y* is an *A-list*, and *z* is a nonnull *A-list*.

Many functions that operate on lists have the same basic structure. For example

$$\textbf{def rec } sum\ x\ =$$
$$\textbf{if } null\ x$$
$$\textbf{then } 0$$
$$\textbf{else } (h\ x) + sum(t\ x)$$

$$sum(1, 2, 3, 4) = 10$$

$$\textbf{def rec } product\ x\ =$$
$$\textbf{if } null\ x$$
$$\textbf{then } 1$$
$$\textbf{else } (h\ x) \times product(t\ x)$$

$$product(1, 2, 3, 4) = 24$$

$$\textbf{def rec } append\ x\ y\ =$$
$$\textbf{if } null\ x$$
$$\textbf{then } y$$
$$\textbf{else } (h\ x){:}(append(t\ x)y)$$

$$append(1, 2)(3, 4, 5) = 1, 2, 3, 4, 5$$

$$\textbf{def rec } concat\ x\ =$$
$$\textbf{if } null\ x$$
$$\textbf{then } ()$$
$$\textbf{else } append(h\ x)(concat(t\ x))$$

$$concat((1, 2), (3, 4), ()) = 1, 2, 3, 4$$

$$\textbf{def rec } map\ f\ x\ =$$
$$\textbf{if } null\ x$$
$$\textbf{then } ()$$
$$\textbf{else } (f(h\ x)){:}(map\ f(t\ x))$$

$$map\ square\ (1, 2, 3, 4) = 1, 4, 9, 16$$

It can be seen that these functions differ only in the first arm of the conditional expression and in the function that combines whatever is produced from the head of the list (usually the head itself) with whatever is produced by applying the same function to the tail of the list. The common parts of these functions can be expressed as a function in which the parts that are not common have been made variables. This function, *list* 1 is defined as:

> **def rec** *list* 1 *a g f x* =
>
>   **if** *null x*
>
>   **then** *a*
>
>   **else** $g(f(h\ x))(list1\ a\ g\ f(t\ x))$

The result of applying the function (*list* 1 *a g f*) to a list is *a* if the list is empty; otherwise it is the result of applying g to two arguments, which are: 1) the result of applying *f* to the head of the list, and 2) the result of applying the same function to the tail of the list. The five functions above can now be redefined in terms of *list* 1. It can be seen that this technique both saves writing and is more likely to produce a correct program because the complex programming (i.e., the conditional expression and looping) has been written once and for all in the function *list* 1. In the following definitions, *I* is the identity function, *K x y* = *x,* and *postfix* adds a new item to the end of a list; so that:

> **def** *postfix x y* = *append y* (*u x*)

The new definitions are:

> **def** *sum* = *list* 1 0 *plus I*
>
> **def** *product* = *list* 1 1 *mult I*
>
> **def** *append x y* = *list* 1 *y prefix I x*
>
> **def** *concat* = *list* 1 () *append I*
>
> **def** *map f* = *list* 1 () *prefix f*

Some other examples are:

> **def** *length* = *list* 1 0 *plus* (*K* 1)
>
> **def** *sumsquares* = *list* 1 0 *plus square*
>
> **def** *reverse* = *list* 1 () *postfix I*
>
> **def** *identity* = *list* 1 () *prefix I*

The types of the arguments of *list* 1 can be deduced from their pattern of application as follows. The whole function (*list* 1 *a g f*) must operate on a list. Suppose that the function is an *A-list* and that it produces a member of *B*; i.e.,

$$(list1\ a\ g\ f)\ \varepsilon\ (A\text{-}list \rightarrow B)$$

then the argument *a* must be a *B*. If the function *f* operates on an *A* and produces a *C*,

$$f\ \varepsilon\ (A \rightarrow C);$$

then g must operate on a C, and the result of applying ($list\ 1\ a\ g\ f$) to an A-list, which is a B, must produce a B. The type of g is therefore

$$g\ \varepsilon\ (C \rightarrow (B \rightarrow B)).$$

The function $list\ 1$ has the property

$$list1\ a\ g\ f(append\ x\ y) = list1(list1\ a\ g\ f\ y)g\ f\ x$$

provided that the types of a, g, and f have the types above for some sets A, B, and C, and that x and y are both A-lists. This may be proved in a mechanical fashion from the axioms for lists and the definitions of $list\ 1$ and *append*. The property emphasizes the fact that in all functions defined in terms of $list\ 1$, the list is first scanned to its end, and then the actions g are taken on the way back. The same result can be obtained in two ways: 1) by concatenating two lists x and y, and then applying ($list\ 1\ a\ g\ f$) to the result, and 2) by first applying ($list\ 1\ a\ g\ f$) to the second list (y) and then using the result as an initial value (the a argument) for the same function when it is applied to the first list (x).

There are two cases to be examined to prove this property of $list\ 1$: 1) when x is the null list, and 2) when x is nonnull. The proof proceeds by reducing each side to the same expression. Let

$$\text{LHS} = list1\ a\ g\ f(list1\ y\ prefix\ I\ x)$$
$$\text{RHS}\ list1\ (list1\ a\ g\ f\ y)g\ f\ x.$$

If x is the null list,

$$\text{LHS} = list1\ a\ g\ f(list1\ y\ prefix\ I\ ())$$
$$= list1\ a\ g\ f\ y$$

and

$$\text{RHS} = list1\ (list1\ a\ g\ f\ y)g\ f\ ()$$
$$= list1\ a\ g\ f\ y$$

Both steps are applications of the rule: $list\ 1\ a\ g\ f() = a$.

When x is not null it takes the form $h\ x{:}t\ x$, and so

$$\text{LHS} = list1\ a\ g\ f(list1\ y\ prefix\ I\ (h\ x{:}t\ x))$$
$$= list1\ a\ g\ f(prefix\ (I\ (h\ x))(list1\ y\ prefix\ I\ (t\ x)))$$
$$= list1\ a\ g\ f(prefix\ (h\ x)(list1\ y\ prefix\ I\ (t\ x)))$$
$$= g(f(h\ x))(list1\ a\ g\ f(list1\ y\ prefix\ I\ (t\ x)))$$
$$= g(f(h\ x))(list1\ (list1\ a\ g\ f\ y)g\ f(t\ x))$$

and

$$\text{RHS} = list1 \ (list1 \ a \ g \ f \ y) \ g \ f (h \ x:t \ x)$$
$$= g(f(h \ x))(list1 \ (list1 \ a \ g \ f \ y) g \ f (t \ x))$$
$$= \text{LHS}.$$

Each step is an application of the rule,

$$list1 \ a \ g \ f(h \ x:t \ x) = g(f(h \ x))(list1 \ a \ g \ f(t \ x)),$$

except the last in the reduction of the LHS in which the property to be proved is assumed to be true for the tail of the list. All the other properties of functions that follow can be proved in the same mechanical fashion.

The functions defined in terms of $list \ 1$ all scan the list (i.e., accumulate the result) from right to left. There is a second family of functions which scan lists from left to right. Such a group can be defined by using a function called $list \ 2$, defined as:

> **def** $list2 \ a \ g \ f \ x =$
>> **if** $null \ x$
>> **then** a
>> **else** $list2(g \ (f(h \ x))a) g \ f (t \ x)$

The arguments $a, \ g, \ f$ and x must be the same types as those for $list \ 1$. The function $list2$ has a property analogous to that of $list \ 1$, namely:

$$list2 \ a \ g \ f(append \ x \ y) = list2 \ (list2 \ a \ g \ f \ x) g \ f \ y.$$

There are two ways of producing the same result: 1) by appending two lists x and y and then applying $(list \ 2 \ a \ g \ f)$ to the result, or 2) by first applying the function to x and then using the result as an initial value for the same function when it is applied to y. Note that the roles of x and y are reversed in the properties of $list \ 1$ and $list \ 2$. In the case of functions defined using $list \ 2$, the result is accumulated as the list is scanned from left to right. The function $list \ 2$ can be implemented by the *iterative* program below, and so it may be more efficient to use $list \ 2$ than $list \ 1$.

> $L:$ **if** $null \ x$
>> **then** a
>> **else**
>>> $a := g(f(h \ x))a$
>>> $x := t \ x$
>>> **go to** L

It is also possible to prove that

$$list1 \; a \; g \; f \; x \; = \; list2 \; a \; g \; f (reverse \; x).$$

This means that ($list \; 1 \; a \; g \; f$) produces the same result when applied to a list x as ($list2 \; a \; g \; f$) does when it is applied to the reversal of the list x. Since

$$append \; x \; y \; = \; list1 \; y \; prefix \; x \; = \; list2 \; x \; postfix \; y$$

and

$$reverse \; = \; list1 \; () \; postfix \; I \; = \; list2 \; () \; prefix \; I$$

it is possible, by substituting particular values for a, g and f in the relationship between $list \; 1$ and $list \; 2$, to prove that

$$reverse(reverse \; x) \; = \; x$$
$$append(append \; x \; y)z \; = \; append \; x(append \; y \; z)$$
$$reverse(append \; x \; y) \; = \; append \; (reverse \; y)(reverse \; x).$$

The generating function for an A-list can be obtained immediately from the structure description for lists

$$list \; A \; = \; du(cp(), \; cp(A, \; list \; A))$$
$$list(x) \; = \; 1 \; + \; x \; list(x)$$

There is one list having no components (x^0 or 1), and a nonnull A-list which has a *head* which is an $A(x)$, and ($\times$) a tail which is an A-list (list(x)). This can be solved, giving

$$list(x) \; = \; 1/(1-x)$$
$$= \; 1 \; + \; x \; + \; x^2 \; + \; x^3 \; + \; x^4 \; + \; ...$$

Another way to read the definition is: a list either is empty or is a 1-list or is a 2-list, etc. The generating function for a list of length n is x^n. If the components of a list are either x or y, ($x \; + \; y$), then the number of lists containing m x's is the coefficient of x^m in ($x \; + \; y$)^n. This is the binomial coefficient

$$\binom{n}{m} \; = \; \frac{n!}{m! \; (n-m)!}$$

which is also the number of possible choices of m different objects from a set of size n.

The generating functions for lists whose lengths are restricted in some way can be written down immediately. For instance the nonnull lists have generating function

$$nlist(x) \; = \; x/(1-x) \; = \; x \; + \; x^2 \; + \; x^3 \; + \; x^4 \; + \; ....$$

As another example, the generating function for all lists no longer than j is

$$(1 - x^{j+1})/(1 - x) = 1 + x + x^2 + \ldots + x^j.$$

Also, if only 0, 3, and 5-lists are allowed, then the generating function is $1 + x^3 + x^5$. If a different object is associated with each list position (suppose that x_j is associated with position j) and either occupies that position or not, then the resulting generating function is

$$(1 + x_1 x)(1 + x_2 x) \ldots (1 + x_n x).$$

The coefficient of x^m contains all combinations of products of m different x_j's drawn from $x_1, x_2, x_3, \ldots, x_n$:

$$(1 + x_1 x)(1 + x_2 x)(1 + x_3 x) =$$
$$1 + (x_1 + x_2 + x_3) + (x_1 x_2 + x_1 x_3 + x_2 x_3)x^2 + x_1 x_2 x_3 x^3,$$

and are called elementary symmetric functions.

The generating function for a list of length n whose j^{th} item has generating function $A_j(x)$ is the product $\Pi A_j(x)$. In particular, if each A_j is a list of x_j's, the resulting generating function is that for a list of lists,

$$[1/(1 - x_1 x)][1/(1 - x_2 x)][1/(1 - x_3 x)] \ldots [1/(1 - x_n x)],$$

in which the coefficient of x^m is all combinations of $x_1, x_2, \ldots, x_n$ of size m with repetitions allowed. If all x_j are set equal to 1, the result is the generating function for the number of such combinations.

$$1/(1 - x)^n = \sum_{m \geq 0} \binom{m+n-1}{m} x^m$$

Since the generating function for nonnull lists is $x/(1 - x)$, the same problem in which at least one of each x_j must appear in each combination has generating function

$$x/(1 - x)^n = \sum_{m \geq n} \binom{m-1}{m-n} x^m$$

The generating function that corresponds to the structure $du(x_1, x_2, \ldots, x_n)$ is $x_1 + x_2 + \ldots x_n$, and so the generating function for the number of n-lists in which one of $x_1, x_2, \ldots, x_n$ can occur in each position is

$$1/[1 - (x_1 + x_2 + \ldots + x_n)x]$$

It is possible to create new generating functions by substituting a generating function for a variable in another generating function. The method can be illustrated as follows. The generating function for a nonempty list of nonempty lists is nlist(nlist(x)). Where nlist $(x) = x/(1 - x)$, it is therefore $x/(1 - 2x)$, and there are 2^{n-1} configurations with n atomic components. The structures are usually called compositions. The 8 compositions of the number 4 are given in Fig. 3.1.

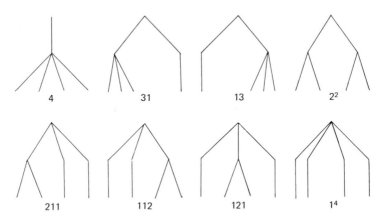

Fig. 3.1   Compositions of 4

The function associated with nlist that scans a nonempty list is

$$\textbf{def } nlist\ g\ f\ x\ =$$
$$\textbf{if } null\ (t\ x)$$
$$\textbf{then } f(x)$$
$$\textbf{else } g(f(h\ x))(nlist\ g\ f(t\ x))$$

and the function for scanning compositions is $nlist\ g_1\ (nlist\ g_2\ f)$. A composition of $n + 1$ can be obtained from a composition of n by either adding a new head to the list or by adding a new head to the head of the list.

A function that operates on two equal length lists and that first combines corresponding members by using a function f and then combines the results by using successive applications of a function g is defined as:

$$\textbf{def rec } zip\ a\ g\ f\ x\ y\ =$$
$$\textbf{if } null\ x$$
$$\textbf{then } a$$
$$\textbf{else } g(f(h\ x)(h\ y))(zip\ a\ g\ f(t\ x)(t\ y))$$

Such a function can be used, for example, to form the scalar product (zip 0 $plus\ mult$) or to produce a list of pairs from a pair of lists by:

$$\textbf{def } zips\ =\ zip\ ()\ prefix\ pair$$

Another way of providing the same family of functions is to first produce a list of pairs and then use a *list* 1 or *list* 2 function. It follows that

$$zip\ a\ g\ f\ x\ y\ =\ list1\ a\ g\ f1\ (zips\ x\ y)$$
$$\textbf{where}\ f1(x, y) = f\ x\ y.$$

3.5 LIST STRUCTURES

An A-list structure is either atomic, and is an A, or is an (A-list structure)-list. Thus:

$$\textbf{def rec}\ liststructure\ A\ =\ du(A, list(liststructure\ A))$$
$$\textbf{def}\ atomic, nonatomic\ =\ predicates\ (liststructure\ A)$$

The functions *list* 1 and *list* 2 can be extended to list structures as follows:

$$\textbf{def rec}\ ls1\ a\ g\ f\ x\ =$$
$$\quad \textbf{if}\ atomic\ x$$
$$\quad \textbf{then}\ f\ x$$
$$\quad \textbf{else}\ list1\ a\ g\ (ls1\ a\ g\ f)\ x$$
$$\textbf{def rec}\ ls2\ a\ g\ f\ x\ =$$
$$\quad \textbf{if}\ atomic\ x$$
$$\quad \textbf{then}\ f\ x$$
$$\quad \textbf{else}\ list2\ a\ g\ (ls2\ a\ g\ f)\ x$$

In both, the same function, either $(ls1\ a\ g\ f)$ or $(ls2\ a\ g\ f)$, is applied to a component list structure as is applied to the list structure itself. In this case f and $(ls1\ a\ g\ f)$ must produce the same type of object. It follows that these functions and some of the functions defined for lists (i.e., those with $g\ \varepsilon\ B \rightarrow (B \rightarrow B)$ form some B) can be extended to be applicable to list structures. For example

$$\textbf{def}\ sumls\ =\ ls1\ 0\ plus\ I\ =\ ls2\ 0\ plus\ I$$
$$\textbf{def}\ productsls\ =\ ls1\ 1\ mult\ I\ =\ ls2\ 1\ mult\ I$$
$$\textbf{def}\ concatls\ =\ ls1\ ()\ append\ u$$
$$\textbf{def}\ mapls\ =\ ls1\ ()\ prefix\ f$$
$$\textbf{def}\ lengthls\ =\ ls1\ 0\ plus\ (K\ 1)$$
$$\textbf{def}\ sumsquaresls\ =\ ls1\ 0\ plus\ square$$
$$\textbf{def}\ reversels\ =\ ls1\ ()\ postfix\ I$$
$$\textbf{def}\ identityls\ =\ ls1\ ()\ prefix\ I$$

The first two functions take the sum and the product of the atomic components of a list structure; *concatls* flattens the list structure into a list; *(mapls f)* applies *f* to all the atomic components and produces a list structure which contains the results and has the same shape as the original; *lengthls* counts the number of atomic components; and *reversels* reverses all the lists in the structure. Examples of the applications of these functions follow:

> **let** *ls* = (1, (2, (3, 4)), 5)
>
> *sumls ls* = 15
>
> *productls ls* = 120
>
> *concatls ls* = 1, 2, 3, 4, 5
>
> *mapls ls* = (1, (4, (9, 16)), 25)
>
> *lengthls ls* = 5
>
> *sumsquaresls ls* = 55
>
> *reversels ls* = (5, ((4, 3), 2), 1)
>
> *identityls ls* = (1, (2, (3, 4)), 5)

The relation between *list* 1 and *list* 2 is extended to

$$ls1\ a\ g\ f\ x = ls2\ a\ g\ f\ (reversels\ x).$$

The generating function for a list structure is given by

$$\text{ls}(x) = x + \text{list (ls } x).$$

There is an infinite number of list structures containing n atomic elements, because 0-lists and 1-lists can be added to extend the structure without adding any atomic components. If 0-lists and 1-lists are disallowed, the list-generating function becomes $x^2/(1 - x)$ and the generating function for list structures without 0-lists or 1-lists is:

$$\text{nls}(x) = x + \text{nls}^2(x)/[1 - \text{nls}(x)]$$

which can be solved to give

$$2\text{nls}^2(x) - (1 + x)\text{nls}(x) + x = 0$$

$$\text{nls}(x) = 1/4\left(1 + x - \sqrt{1 - 6x + x^2}\right)$$

$$= x + x^2 + 3x^3 + 11x^4 + 45x^5 + \ldots$$

This generating function enumerates the ways of bracketing a sum of n numbers, in which each bracket must include at least two expressions.

Four instances are

$$a$$

$$a + b$$

$$a + b + c, (a + b) + c, a + (b + c)$$

$$a + b + c + d, (a + b + c) + d, a + (b + c + d), (a + b)$$
$$+ c + d, a + (b + c) + d, a + b + (c + d), (a + b) + (c + d),$$
$$((a + b) + c) + d, (a + (b + c)) + d, a + ((b + c) + d), a$$
$$+ (b + (c + d))$$

This structure can be rearranged by tipping up the list structure so that the leftmost edge becomes the top level of the structure. The generating function can be rearranged as

$$2\mathrm{nls}^2(x) - (1 + x)\mathrm{nls}(x) + x = 0$$
$$[2\mathrm{nls}(x) - 1]\mathrm{nls}(x) + x[1 - \mathrm{nls}(x)] = 0$$
$$\mathrm{nls}(x) = x/\{\, 1 - [\mathrm{nls}(x)/1 - \mathrm{nls}(x)]\}$$

which corresponds to the structure:

- An A-nls

  has a *root* which is an A

  and a *body* which is an ((A-nls)-nlist)-list.

3.6   TREES AND FORESTS

List structures are ordered, rooted trees with components only at their leaves or end points. Another structure that is often useful is a tree having components at every node. If the interior and end point are to be treated in the same way, the relevant tree structure is defined as follows.

- An A-tree has a *root* which is an A

      and a *listing* which is an A-forest.

  An A-forest is an (A-tree)-list.

The function used to construct a tree from an A and an A-forest will be called *ctree* and is defined:

$$\textbf{def rec } tree\ A\ =\ cp(A, forest\ A)$$
$$\textbf{and } forest\ A\ =\ list(tree\ A)$$
$$\textbf{def } root,\ listing\ =\ selectors\ (tree\ A)$$
$$\textbf{def } ctree\ =\ construct\ (tree\ A)$$

Two families of functions can be defined using *list* 1 and *list* 2:

def rec *tree*1 *a g f x* = *f*(*root x*)(*forest*1 *a g f*(*listing x*))

and *forest*1 *a g f x* = *list*1 *a g* (*tree*1 *a g f*) *x*

def rec *tree*2 *a g f x* = *f*(*root x*)(*forest*2 *a g f*(*listing x*))

and *forest*2 *a g f* = *list*2 *a g* (*tree*2 *a g f*)

In this representation, an end point is a tree whose listing is the null list. As in the case of list structures, the same function is applied to both a component tree and the whole tree or to both a component forest and the whole forest.

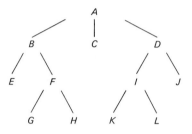

Fig. 3.2   An example of a tree

The expression below will be used to describe the tree in Fig. 3.2. The comma is used for lists and the infixed semicolon is used for *ctree*.

$$A; (B; (E; (), F; (G; (), H; ())), C; (), D; (I; (K; (), L; ()), J; ()))$$

There are several ways of scanning the nodes of a tree. They can be expressed as functions for listing all the atomic components. For instance, the result of applying *preorder* = (*tree*1 () *append prefix*) to the tree in Fig. 3.3 is

$$A \ B \ E \ F \ G \ H \ C \ D \ I \ K \ L \ J;$$

the result of applying *postorder* = (*tree*1 () *append postfix*) is

$$E \ G \ H \ F \ B \ C \ K \ L \ I \ J \ D \ A;$$

and (*tree*2 () *append prefix*) produces

$$A \ D \ J \ I \ L \ K \ C \ B \ F \ H \ G \ E,$$

which is also the result of applying preorder to the reversed tree. In a similar way (*tree*2 () *append postfix*) produces the same result as applying *postorder*

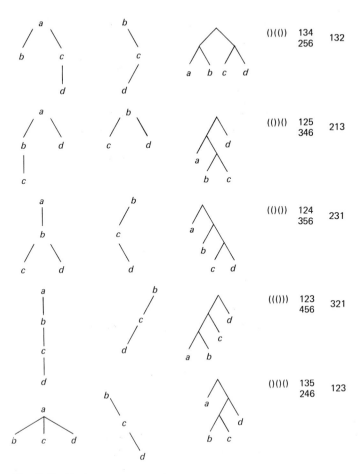

Fig. 3.3 Various correspondences

to the reversed tree. A tree and all its lists can be reversed by using:

$$\textbf{def } reversetree \;=\; tree2\;() \;prefix\; ctree$$

while a forest can be reversed by:

$$\textbf{def } reverseforest \;=\; forest2\;() \;prefix\; ctree.$$

It is easy to prove that

$$reversetree(reversetree\; x) \;=\; x$$

and

$$reverseforest(reverseforest\; x) \;=\; x$$

Each function on a tree which is defined in terms of *tree* 1 or *tree* 2 can be applied to a forest by using the same arguments and the functions *forest* 1 and *forest* 2, respectively, and vice versa. The relationship between *list* 1 and *list* 2 may be used to give the following relationships between *tree* 1 and *tree* 2 and *forest* 1 and *forest* 2:

$$tree1 \ a \ g f x \ = \ tree2 \ a \ g f (reversetree \ x)$$
$$forest1 \ a \ g f x \ = \ forest2 \ a \ g f \ (reverseforest \ x).$$

The generating functions for trees and forests are:

$$tree(x) \ = \ x \ forest(x)$$
$$forest \ x \ = \ list(tree \ x)$$

therefore

$$tree(x) = x \ list[tree(x)] = x/[1 - tree(x)]$$
$$tree^2 (x) - tree(x) + x = 0$$
$$tree(x) = 1/2(1 - \sqrt{1 - 4x})$$
$$= \Sigma \ 1/n \binom{2n-2}{n-1}x$$
$$= x + x^2 + 2x^3 + 5x^4 + 14x^5 + 42x^6 + 132x^7 + \dots .$$

Also since

$$tree(x) \ = \ x \ forest(x)$$

then

$$x \ forest^2 (x) - forest(x) + 1 = 0.$$

The forests with n nodes are in correspondence with the trees with $n + 1$ nodes formed by adding a root. The forests with p nodes are in $1 - 1$ correspondence with the set of *lattice permutations* of the collection of p opening and p closing parentheses. A lattice permutation of a collection of $n_1 \ x_1$'s, $n_2 \ x_2$'s, ..., $n_m \ x_m$'s is a permutation in which any initial segment contains no more x_{i-1}'s than x_i's, for all $1 \le i \le m$. Since there are an equal number of open and closing parentheses, and since the opening one always precedes a closing one, the strings all have the property of lattice permutations that there are a greater number of opening than closing parentheses in any initial segment of a string. The brackets can be interpreted as instructions for adding an item to, and removing an item from, a pushdown list. The set of lattice permutations therefore corresponds to all possible ways of adding and removing items from a pushdown list.

Alternatively the string of brackets can be interpreted as instructions for constructing a two row array. The instruction "(" adds its rank to the first row while ")" adds its rank to the second. At each stage the first row is never shorter than the second. The set of lattice permutations with p opening and p closing brackets are therefore in correspondence with all the p-column 2-row arrays filled with the numbers 1 to $2p$ in such a way that there is an ascending order in both the rows and the columns. These correspondences for $p = 3$ are illustrated in Fig. 3.3.

Each forest defines a permutation produced by applying its bracket string to the list 1, 2, 3, 4, ... The opening bracket is interpreted as an instruction for loading the next number to the pushdown list. A closing bracket is interpreted as an instruction which takes the number from the top of the pushdown list and postfixes it to the output permutation. The permutation produced can also be obtained by labeling the forest with the numbers 1, 2, 3, 4, ..., so that 1, 2, 3, 4, ... is produced by *preorder,* and then by applying *postorder* to this tree. These correspondences are shown in Fig. 3.4.

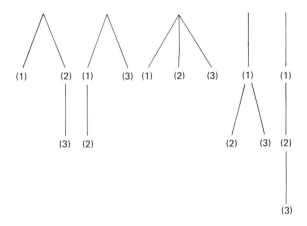

Fig. 3.4   Lattice permutations

Another way to exhibit the structure of a tree, ignoring the components, is to replace each root by the length of its listing and then list these lengths in preorder. The first tree in Fig. 3.5, for example, would be described by the string 320200022000. There is a correspondence between forests which is such that: a) the degree of each root is the same in both and b) that the same list of roots is produced by scanning the first level by level as is produced by applying *preorder* to the second. This can be done by adding the length of the listing to each root and then by scanning the tree level

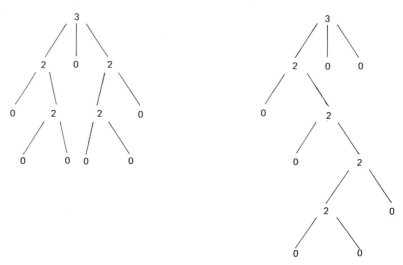

Fig. 3.5. A correspondence which preserves the degrees of the nodes

by level. The resulting code word must describe a forest because a root is visited before any of its subtrees. An example of this correspondence is shown in Fig. 3.5.

3.7 BINARY TREES

Two more restricted types of trees are considered next. We will show that correspondences can be set up between both of these types of trees and the trees just defined. The first type is called a *binary tree* and can have 0 or 2 subtrees.

- An A-binary tree

    either is *empty*

    or has a *root* which is an A

    and a *left* and a *right,* both of which are A-binary trees.

$$\textbf{def rec } btree\ A\ =\ du(cp(),\ cp(A,\ btree\ A,\ btree\ A))$$
$$\textbf{def } empty,\ nonempty\ =\ predicates\ (btree\ A\ )$$
$$\textbf{def } emptyc,\ nonemptyc\ =\ parts\ (btree\ A\ )$$
$$\textbf{def } root,\ left,\ right\ =\ selectors\ nonemptyc$$
$$\textbf{def } cbtree\ =\ construct\ nonemptyc$$
$$\textbf{def } etree\ =\ construct\ emptyc\ ()$$

The function for constructing nonempty trees is called *cbtree.* The name of the empty tree is *etree*. The family of functions that scan binary trees can be produced by using

> **def rec** *btree*1 *a g f x* =
>> **if** *empty x*
>> **then** *a*
>> **else** $g(f(root\ x))$
>>> $(btree1\ a\ g\ f(left\ x))$
>>> $(btree1\ a\ g\ f(right\ x))$

The types of the arguments can be deduced as follows; if

$$(btree1\ a\ g\ f)\ \varepsilon\ A\text{-}binary\ tree \rightarrow B,$$

then

$$a\ \varepsilon\ B,\ f\ \varepsilon\ A \rightarrow C,$$

and

$$g\ \varepsilon\ C \rightarrow (B \rightarrow (B \rightarrow B)).$$

It is possible to define another function on binary trees which interchanges the roles of the left and right subtrees, called *btree*2.

> **def rec** *btree*2 *a g f x* =
>> **if** *empty x*
>> **then** *a*
>> **else** $g(f(root\ x))$
>>> $(btree2\ a\ g\ f(right\ x))$
>>> $(btree2\ a\ g\ f(left\ x))$

The function for reversing a binary tree is:

> **def** *reversebtree* = $(btree2\ etree\ cbtree\ I)$
>> = $(btree1\ etree\ (C\ ctree)\ I)$
> **where** $C f x y = f y x.$

There is a correspondence between a binary tree and a forest which can perhaps best be seen by unwinding, so to speak, the structure definition of a forest.

- An A-forest is an (A-tree)-list
  > i.e., either is *null*
  >> or has a *h* which is an A-tree
  >> and a *t* which is an (A-tree)-list.

In other words:

- An A-forest either is *null*

$\qquad$ or has a (*root* • *h*) which is an A

$\qquad$ and a (*listing* • *h*) which is an A-forest

$\qquad$ and a *t* which is also an A-forest.

This definition has the same structure as a binary tree, and the correspondence can be expressed as the following correspondence between the functions on binary trees and forests.

Binary tree	Forest
empty	*null*
etree	()
root	*root* • *h*
left	*listing* • *h*
right	*t*
btree x y z	(*ctree x y*):*z*

Every function that either operates on or constructs binary trees can be transformed to a function that operates on or constructs forests. An example of the correspondence between forests and binary trees is shown in Fig. 3.6. There is a transformation of forests which corresponds to a reversal of the corresponding binary tree. It is clear that the *forest* 1 function can be "unwound" in the same way as the structure definition to give:

$$
\begin{aligned}
forest1\ a\ g\ f\ x\ =\ &list1\ a\ g\ (tree1\ a\ g\ f)x \\
=\ &\textbf{if } null\ x \\
&\textbf{then } a \\
&\textbf{else } g(tree1\ a\ g\ f(h\ x)) \\
&\qquad\qquad (list1\ a\ g(tree1\ a\ g\ f(t\ x))) \\
=\ &\textbf{if } null\ x \\
&\textbf{then } a \\
&\textbf{else } g(f(root(h\ x)) \\
&\qquad\qquad (forest1\ a\ g\ f(listing(h\ x)))) \\
&\qquad (forest1\ a\ g\ f(t\ x))
\end{aligned}
$$

Another scanning function for forests called *forest* 3 can be defined by interchanging the rôles of the two forests in positions *listing* • *h* and *t* in this

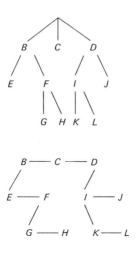

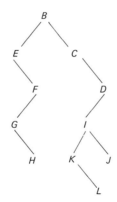

Fig. 3.6 A correspondence between a forest and a binary tree

definition, giving

> **def rec** *forest3 a g f x =*
>> **if** *null x*
>> **then** *a*
>> **else** *g*(*f*(*root*(*h x*))
>>>> (*forest3 a g f*(*t x*)))
>>>> (*forest3 a g f*(*listing*(*h x*))))

The function for rotating a forest, i.e., producing the forest corresponding to the reversal of the binary tree underlying the original, is

$$\textbf{def } rotate = forest3 \ () \ ctree \ prefix$$

It follows that

$$forest1 \ a \ g \ f \ x = forest3 \ a \ g \ f(rotate \ x)$$
$$rotate(rotate \ x) = x$$

The endorder scan described by Knuth [3–11] is the postorder scan of the rotated forest.

$$\textbf{def } endorder \ x = postorder(rotate \ x)$$
$$= forest1 \ () \ append \ postfix \ (rotate \ x)$$
$$= forest3 \ () \ append \ postfix \ x$$

If *forest2* is given the same treatment as *forest1* then the function *forest4* is produced having the properties:

$$forest2 \ a \ g \ f \ x = forest4 \ a \ g \ f(rotate \ x)$$
$$forest1 \ a \ g \ f \ x = forest4 \ a \ g \ f(rotate(reverse \ x))$$

The rotate operation interchanges the two forests in Fig. 3.7.

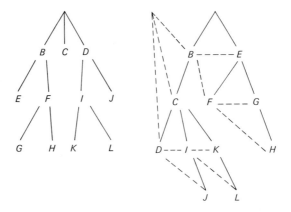

Fig. 3.7   Rotating a forest

A forest can be transformed to a binary tree by the function

$$forest1 \ etree \ g \ pair$$
$$\textbf{where } g(x, y)z = cbtree \ x \ y \ z$$
$$\textbf{and } pair \ x \ y = x, y.$$

A binary tree can be transformed to a forest by

$$btree1\ ()\ g\ I$$

where

$$g\ x\ y\ z\ =\ (ctree\ x\ y){:}z$$

The generating function for binary trees is

$$\text{btree}(x)\ =\ 1\ +\ x\ \text{btree}^2\ (x)$$

and therefore is equal to forest(x).

3.8 COMBINATIONS

Another correspondence can be set up between trees and a structure which is called a *combination*. Combinations are trees with components only at their end points, and each tree can have either 0 or 2 subtrees.

■   An A-combination

   either is *atomic* and is an A

   or has a *left* and a *right,* both of which are A-combinations.

$$\textbf{def rec}\ comb\ A\ =\ du(A,\ cp(comb\ A,\ comb\ A\ ))$$
$$\textbf{def}\ atomic,\ nonatomic\ =\ predicates\ (comb\ A\ )$$
$$\textbf{def}\ atomicc,\ nonatomicc\ =\ parts(comb\ A\ )$$
$$\textbf{def}\ left,\ right\ =\ selectors\ nonatomicc$$
$$\textbf{def}\ combine\ =\ construct\ nonatomicc$$
$$\textbf{def rec}\ comb1\ g\ f\ x\ =$$
$$\quad \textbf{if}\ atomic\ x$$
$$\quad \textbf{then}\ f\ x$$
$$\quad \textbf{else}\ g(comb1\ g\ f\ (left\ x))$$
$$\quad\quad (comb1\ g\ f(right\ x))$$

and *comb2* is defined similarly by interchanging *left* and *right*.

   The interior nodes of a combination with n nodes form a binary tree with $n - 1$ nodes. Both combinations and binary trees can be considered to be special cases of a structure in which the interior and end points are treated as two different types of atomic component.

■   An A-B-binary tree either

   is *atomic* and is a B

   or has a *root* which is an A,

      and a *left* and a *right*

      both of which are A-B-binary trees.

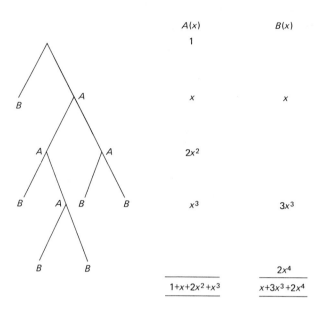

Fig. 3.8   Generating functions for the weights of trees

The binary tree is formed by setting B to empty; the combination is formed by setting A to empty.

The relationships between internal and external nodes can be established by using generating functions. Suppose that $A(x) = \Sigma\, a_k\, x^k$ and $B(x) = \Sigma\, b_k\, x^k$ are generating functions for an A-B-binary tree in which a_k is the number of interior nodes or roots on the kth level and b_k is the number of leaves or atomic components on the kth level. The generating functions for a tree are shown in Fig. 3.8. Then $A(x)$ and $B(x)$ are related as follows

$$A(x) + B(x) = 1 + 2xA(x)$$

The number of leaves is $B(1)$, and the number of interior nodes is $A(1)$. It follows that:

$$B(1) = 1 + A(1), B(1/2) = 1.$$

If $B(x)$ is differentiated with respect to x, giving $B^x\,(x)$, then $B^x\,(1)$ is the sum of the path lengths to the leaves. Also $A^x\,(1)$ is the sum of the path lengths to interior nodes. If the relation between $A(x)$ and $B(x)$ is differentiated then

$$A^x\,(x) + B^x\,(x) = 2A(x) + 2xA^x\,(x)$$

is produced, therefore $B^x\,(1) = 2A(1) + A^x\,(1)$. This means that in an

A-B-binary tree the exterior path length sum B^x (1) is equal to the interior path length sum, A^x (1) plus twice the number of interior nodes.

There is a correspondence between trees and combinations. A function, *tree*5, can be written that scans the tree as if it were the corresponding combination. In this correspondence, a tree with a null listing corresponds to an atomic combination. When nonnull, the tree at the head of the listing corresponds to the *left* of the combination and the tree formed from the root and the tail of the listing corresponds to the *right* of the combination.

> **def** *tree*5 *g f x* =
>> **if** *null*(*listing x*)
>> **then** *f* (*root x*)
>> **else** *g* (*tree*5 *g f* (*h*(*listing x*)))
>>> (*tree*5 *g f* (*ctree*(*root x*)(*t* (*listing x*))))

A tree can be transformed to a combination by using (*tree*5 *combine I*) and a combination to a tree by using

> *comb*1 *g u*
> **where** *g x y* = *ctree*(*root y*)(*x*:*listing y*)
> **and** *u x* = *x*:()

An example of this correspondence is given in Fig. 3.9, in which both trees have been reversed. The correspondence can be interpreted as a translation from a functional notation in which arguments are listed as

$$A(B, C, D(E, F(G, H)))$$

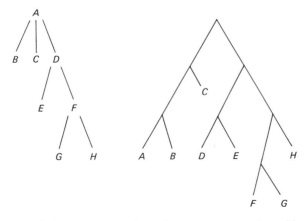

Fig. 3.9   A correspondence between trees and combinations

to one in which only application of a function to a single argument is used, i.e.

$$((A\ B)\ C)\ ((D\ E)\ ((F\ G)\ H)))$$

The generating function for combinations is

$$\text{combination}(x) = x + \text{combination}^2(x)$$

and so combination(x) has the same recurrence relation as tree(x).

3.9   WEIGHTS OF TREES

The weight of a tree with respect to a certain type of atomic component is the sum of the lengths of the paths from each atomic component to the root. The generating functions for the weights of trees can also be obtained almost automatically from the structure definition.

Suppose that $T(x, y)$ is a generating function for a set of trees in which the coefficient of $x^n\ y^w$ is the number of trees with n nodes and weight w. Then if this set of trees is moved down one level then one must be added to the weight for each atomic component in the tree, or for each x in the generating function. The generating function for the set of trees moved down one level is therefore $T(xy, y)$, assuming x to be the variable that corresponds to the atomic components whose depths are to be counted in the weight. Therefore the generating functions for the weights of trees can be written down directly from their structure descriptions.

Using this principle, the weight-generating functions for trees, binary trees, and combinations are

$$\text{wtree}(x, y) = x/1 - \text{wtree}(xy, y)$$
$$\text{wbtree}(x, y) = 1 + x \times \text{wbtree}^2(xy, y)$$
$$\text{wcombination}(x, y) = x + \text{wcombination}^2(xy, y).$$

The expected weight of a set of trees with n nodes can be obtained by first differentiating $T(x, y)$ with respect to y and then setting y equal to one. This has the effect of multiplying the number of trees with weight w and n nodes by the weight w.
If

$$T(x, y) = \Sigma\ t_{nw}\ x^n\ y^w$$

then

$$T^y(x, y) = \Sigma\ wt_{nw}\ x^n\ y^w$$

and

$$T^y(x, 1) = \Sigma\ (\Sigma\ w\ t_{nw})\ x^n$$

The generating function for the number of trees with n atomic components is $T(x, 1)$. If $T(x, 1) = \Sigma\, t_n\, x^n$ then the expected weight of trees with n nodes is $\Sigma\, wt_{nw}/t_n$. If wcombination is abbreviated to c, then the recurrence relation for combinations can be differentiated with respect to y to give

$$2c(xy, y)[c^y(xy, y) + xc^x(xy, y)] = c^y(x, y),$$

therefore,

$$c^y(x, 1) = 2xc(x, 1)c^x(x, 1)/(1-2c)(x, 1)$$
$$= x\left(1 - \sqrt{1 - 4x}\right)/(1 - 4x),$$

hence the sum of the weights of combinations with n + 1 nodes is $4^n - \binom{2n}{n}$.

By a similar argument the sum of the weights of all the trees with n + 1 nodes or forests with n nodes is $(1/2)(4^n - \binom{2n}{n})$, i.e., just half of that for combinations. The sum of the weights of binary trees with n nodes is $4^n - (3n + 1)/(n + 1)\binom{2n}{n}$. In this case these two last results can be derived from the other by considering the correspondences just described.

3.10 SEQUENCES, COROUTINES, AND STREAMS

When one function produces a list in its natural order and another processes the list items in the same order, it is often unnecessary to produce the whole list before applying the second function to it. The two functions can be combined so that, at any stage, the second function can issue a demand for the next list item which is then provided by the first function. The creation of the next list item is thereby delayed until it is actually needed. It is often easier to write programs in two stages in which the list is an intermediate result of the computation. However, it is more economical in storage to use the combination of the two functions in which only one member of the list appears as an intermediate result. This section contains an examination of methods of combining functions in this way. It is possible to have the best of both worlds by writing the program as if the whole list appeared as an intermediate result, but having the actual implementation create only one member at a time. The function which is called upon to produce the next item must both produce it and reset itself so it is prepared to deliver the remainder, or tail of the list, the next time it is called.

The data structure which is relevant is an A-sequence, defined as follows:

- An A-sequence has a *hs* which is an A

    and a *ts* which is an A-sequence.

Streams. A sequence is therefore an infinite list, and the problem of conserving storage for its representation inside a computer becomes even more pressing. A sequence can be represented by a particular type of function, called a *stream function* or *stream*. A *stream* is applicable to an empty list of arguments and produces a pair whose first is the next item in the sequence and whose second is a stream for the tail of the sequence. Thus

$$A\text{-}stream \subseteq (null\ list \rightarrow A \times A\text{-}stream)$$

The head, tail, and prefix functions for a stream are defined as follows:

$$\textbf{def}\ hs\ s\ =\ first(s())$$
$$\textbf{def}\ ts\ s\ =\ second(s())$$
$$\textbf{def}\ prefixs\ x\ s\ =\ \lambda().x,\ s$$

The *hs* of a stream is the first member of the pair that results from applying the stream to the null list. The *ts* is the second member. It follows that s is applied each time either *hs* or *ts* is applied. It is often more economical to make sure that the stream is only applied once by using a construction such as

$$\textbf{let}\ x,\ y\ =\ s();\ \ldots x \ldots y \ldots x \ldots x \ldots.$$

A stream can be constructed from its head x and its tail s by the function *prefixs* $x\ s\ =\ \lambda().x,\ s$. When applied to the null list, this function produces the pair (x, s). The axioms that relate streams and their components are:

$$hs(\lambda().(x, y))\ =\ x$$
$$ts(\lambda().(x, y))\ =\ y$$
$$prefixs(hs\ z)(ts\ z)\ =\ z$$

Stream processing functions. A number of examples of stream-processing functions which are analogous to list-processing functions are defined below.

Example 1

Given a transformer f and an initial value x, a stream function for the sequence $x, f x, f^2 x, f^3 x \ldots$ may be obtained by using:

$$\textbf{def rec}\ generate\ f\ x\ ()\ =\ x,\ generate\ f(f\ x)$$

The first member of the sequence of the stream (*generate f x*) is x, and the remainder of the sequence is represented by the stream (*generate f* $(f\ x)$). Given zero and a successor function, the sequence of nonnegative integers can be represented by the stream *integer* = (*generate successor* 0) = 0, 1, 2, 3, ...

Example 2

The stream representations of sequences can be treated as if they were lists. It is possible to transform streams to other streams, for instance, by using the function *maps* defined as:

$$\textbf{def rec } \textit{maps } f s \; () = f x, \textit{maps } f y$$

$$\textbf{where } x, y = s().$$

The function *maps* transforms a sequence $x_1, x_2, x_3 \ldots$ *etc.* into the sequence $f x_1, f x_2, f x_3 \ldots$ *etc.* The function *maps* delays the production of the next member of s until the next member of $(maps \, f \, s)$ is required, it then applies the function f to the first member of s to produce the first member of $(maps \, f \, s)$. The stream for the sequence of squares of nonnegative integers, for instance, is $(maps \, square \, integer) = 0, 1, 4, 9, \ldots$.

Example 3

The function *thefirst* which finds the first member of a sequence having the property p and produces it, together with the remaining stream, as a result is defined below

$$\textbf{def rec } \textit{thefirst } p s \; =$$

$$\textbf{let } x, y = s()$$

$$\textbf{if } p \, x$$

$$\textbf{then } x, y$$

$$\textbf{else } \textit{thefirst } p \, y$$

Assuming that the predicate *nonspace* tests whether a character is a nonspace character then the next nonspace character can be obtained from a character-stream by applying (*thefirst nonspace*) to it and selecting the first of the pair produced. As another example, the first integer whose square is greater than 1000 is the first member of the pair

$$\textit{thefirst } p \textit{ integer } \textbf{where } p \, x = x^2 > 1000$$

Example 4

The function *filter* operates on a stream and a predicate p, and produces a stream for those members having the property p. This function is defined as:

$$\textbf{def rec } \textit{filter } p s \; =$$

$$\textbf{let } x, y = s()$$

$$\textbf{if } p \, x$$

$$\textbf{then } \lambda().(x, \textit{filter } p \, y)$$

$$\textbf{else } \textit{filter } p \, y$$

A stream of nonspace characters can then be obtained from a character-stream by applying (*filter nonspace*) to it. As another example, (*filter prime integer*) is the stream of prime numbers.

Example 5

Two streams can be processed to produce a third by a function which is analogous to *zip* 1.

$$\textbf{def rec } zips\ f\ x\ y\ =\ \lambda().(f(hs\ x)(hs\ y)),\ (zips\ f\ (ts\ x)(ts\ y))$$

The stream of pairs is produced from two streams by (*zips pair*).

Example 6

Streams are most useful for implementing functions which process character-streams from input. The function *while* produces a list from the initial segment of a stream as long as all its members have the property *p*.

$$\textbf{def rec } while\ p\ s\ =$$
$$\textbf{let } x,\ y\ =\ s()$$
$$\textbf{if } p\ x$$
$$\textbf{then let } u,\ v\ =\ while\ p\ y$$
$$x:u,\ v$$
$$\textbf{else } (),\ s$$

A related function is *until* $p\ =\ while\ (not \cdot p)$. If the predicate *sameline* is $not \cdot (equal\ newline)$ where *newline* is the carriage-return line-feed character, then the function (*while sameline*) operates on a character stream and produces a pair whose *first* is the next line of input and whose *second* is the remaining stream. To be able to reapply the same function, the newline character must be removed by using $remove(x,\ y)\ =\ x,\ ts\ y$.

Example 7

Any function that produces a pair whose second member is the same type as its argument can be made into a stream-transforming function by applying to it a function called *next*, defined as follows:

$$\textbf{def rec } next\ r\ s\ =$$
$$\textbf{let } x,\ y\ =\ r\ s$$
$$\lambda().x,\ next\ r\ y$$

The function *next* is applicable to any function of the type

$$r\ \varepsilon\ A \rightarrow B \times A,$$

and produces a function of the type

$$A \rightarrow B\text{-}stream.$$

It follows that

$$next \; \varepsilon \; [(A \rightarrow B \times A) \rightarrow (A \rightarrow B\text{-}stream)].$$

The function *filter* can be redefined in terms of *next* as follows:

$$\textbf{def } filter \; p \; s \; = \; next(thefirst \; p)s.$$

The function,

$$\{next[remove \cdot (while \; sameline)]\} \; \textbf{where} \; \; remove(x, y) \; = \; x, \; ts \; y,$$

converts a character stream containing newline characters into a line-stream in which the lines are the character-lists between adjacent newline characters.

Example 8

The inverse operation converts a list-stream into a character-stream. Suppose that *concats* is a function for flattening a line-stream into a character-stream, or more generally, transforming an (A-list)-stream into an A-stream, then:

$$\textbf{def rec } concats \; s \; =$$
$$\textbf{let } x, y \; = \; s()$$
$$\textbf{if } null \; x$$
$$\textbf{then } concats \; y$$
$$\textbf{else } \lambda().(h \; x, \; concats \; \lambda().t \; x, \; y)$$

The inverse operation of putting back the newline characters and flattening the line stream into a character stream is then (*concats* · *maps* (*postfix newline*)). The operation of *concats* is similar to an input-buffering process in which blocks of records are read from outside, but the program requires the individual records one at a time. The function *concats* therefore converts a block-reading routine into a next-record routine.

Representing lists by streams.   To represent a list by a stream, some object which cannot be a list item must be chosen to serve as an indication in the stream for the end of the list. This will be called *end*. In fact, since a stream can always be applied, the stream corresponding to an infinite list of *end*'s will serve as the indication. The null stream can be defined by *nullists* = (*generate I end*). The predicate for the null stream is defined as *nulls x* = ((*hs x*) = *end*). These lists, which are represented as streams, can be treated as if they were lists. The correspondence between the list functions and stream functions is given below

Lists	Streams
null x	*nulls x* = ((*hs x*) = *end*)
h	*hs*
t	*ts*
()	*nullists*
x : *y*	λ().*x*, *y*
prefix	*prefixs*
u	*us x* = *prefixs x nullists*

It follows that any function which operates on, or produces, lists can be immediately transformed to a function which operates on, or produces, streams. General purpose stream functions can be defined by analogy with the *list* 1 or *list* 2 functions.

> **def rec** *stream*1 *a g f s* =
> > **if** *nulls s*
> > **then** *a*
> > **else** *g* (*f* (*hs s*))(*stream*1 *a g f* (*ts s*))
>
> **def rec** *stream*2 *a g f s* =
> > **if** *nulls s*
> > **then** *a*
> > **else** *stream*2(*g* (*f* (*hs s*))*a*) *g f* (*ts s*)

The *stream* 1 function produces the whole list before operating on it; the *stream* 2 function produces items one at a time. The functions on lists can be carried over to streams. The stream versions of the functions *map, append* and *concat* follow:

> **def rec** *mapl f s* =
> > **if** *nulls s*
> > **then** *nullists*
> > **else** λ().*f* (*hs s*), *mapl f* (*ts s*)
>
> **def rec** *appendl x y* =
> > **if** *nulls x*
> > **then** *y*
> > **else** λ().*hs x*, *appendl*(*ts x*)*y*

> **def rec** *concatl s* =
> > **if** *nulls s*
> > **then** *nullists*
> > **else let** *x, y* = *s*()
> > > **if** *null x*
> > > **then** *concatl y*
> > > **else** λ().*h x, concatl*(λ().*t x, y*)

Note that in this representation, the *appendl* function is more efficient than the *append* function which has to scan through the first argument to create its result. Another way to represent a stream is by a list *x* for its initial segment and a stream *y* for the remainder. The list is similar to an input buffer. Such a stream can be constructed by another variation of *append*:

> **def rec** *appendls x y* () =
> > **if** *null x*
> > **then** *y*()
> > **else** *h x, appendls*(*t x*)*y*

The *concatl* function operates on a stream of lists and produces a stream. There are eight variations of the concatenating function produced by changing the top-level list, the second-level list, or the resulting list to streams. The version in which all three are streams will be called *concatss* and is defined below:

> **def rec** *concatss s* =
> > **if** *nulls s*
> > **then** *nullists*
> > **else let** *x, y* = *s*()
> > > **if** *nulls x*
> > > **then** *concatss y*
> > > **else** λ().*hs x, concatss*(λ().*ts x, y*)

Some care must be taken when a stream is produced to make sure that its elements are not really a list in disguise, in other words, to make sure that the stream elements are not materialized too soon. We have assumed that, in the method of evaluation used, the operator and operand parts of an expression are evaluated and the value of the operator is then applied to the value of the operand. We have also assumed that the body of a lambda expression, i.e., the *M* part of an expression λ*x.M,* is only evaluated when the function is applied. If the *appendl* function were put into the lambda convertible form

$$appendl\ x\ y\ =$$
if *null x*
then *y*
else *prefixs*(*hs x*)(*appendl*(*ts x*)*y*)

then the inner expression *appendl*(*ts x*)*y* would be applied when the function *appendl* is applied to *x* and *y*. This would cause the elements of the stream *x* to be materialized, and prefixed, using *prefixs*, to the stream *y*. In the first version of *appendl*, on the other hand, the expression *appendl* (*ts x*)*y* will only be evaluated when the stream *appendl x y* is applied to the null list. The two definitions of *appendl* are lambda convertible and therefore are equivalent. The assumed method of evaluation, however, causes the two functions to behave in different ways. In order to produce the most delayed version of a stream, the construction $\lambda().x, y$ should be used instead of *prefixs x y* and expressions containing *prefixs* should not be used.

Loop control.   Streams are also useful for implementing the sequences of values taken on by a variable in a **do**, **for**, or **while** loop in a programming language. The loop control can be separated from the loop by using streams. This means that the same loop control can be used with two different loops, or that the same loop can be used with two different loop controls. The list of numbers from 0 to *n*, for example, have the stream *whiles*(*less* (*n* + 1))*integer*, where:

def rec *whiles p s* =
let *x, y* = *s*()
if *p x*
then $\lambda().x$, *whiles p y*
else *nullists*

Again there is a companion function *untils p* = *whiles*(*not* • *p*). The stream corresponding to the ALGOL 60 phrase

a **step** *b* **until** *c*

is

$$untils(\lambda x.x - c \times sign(b) > 0)(generate(plus\ b)a).$$

The stream for

1 **step** 1 **until** *n*

is

$$\textbf{def } til\; n \;=\; whiles(less(n + 1))(generate(plus\; 1)1).$$

This separation of the loop control from the loop permits nonnumerical streams to control the looping. The double loop introduced by a piece of program of the form

$$\textbf{for } i := 1 \textbf{ step } 1 \textbf{ until } n \textbf{ do}$$

$$\textbf{for } j := 1 \textbf{ step } 1 \textbf{ until } m \textbf{ do}$$

could be regarded as a loop controlled by a stream of pairs. The function *map,* which replaces each item in a list by its transform under a function, can be extended to apply to two lists. The function *map* 2, defined as

$$\textbf{def } map2\; f\; x\; y \;=\; \textbf{let } g\; z \;=\; map(f z)y$$

$$concat(map\; g\; x)$$

produces a list of the results of applying f to every pair of items, one from one list and the second from the other. The result of applying $(map\,2\; pair)$ to the two lists $(1, 2, 3)$ and $(4, 5)$ is the cross product

$$(1, 4), (1, 5), (2, 4), (2, 5), (3, 4), (3, 5).$$

There is a stream version of this function, produced by replacing *map* by *mapl,* and *concat* by *concatss,* i.e.,

$$\textbf{def } map2l\; f\; x\; y \;=\; \textbf{let } g\; z \;=\; mapl(f z)y$$

$$concatss(mapl\; g\; x)$$

The double **for** loop control above can be regarded as the stream of pairs $(map\,2l\; pair\; (til\; n)(til\; m))$.

Walking through trees.   It is often useful to be able to scan a data structure by using a stream instead of first listing its elements and then scanning the list. A general technique for producing a list from a structure is first to replace all atomic elements by 1-lists, then to produce a list of lists from each nonatomic component, and finally to concatenate these lists. These functions for producing lists can be systematically changed so that each atomic element is converted into a 1-stream, each nonatomic component is converted into a stream of streams, and this stream of streams is then concatenated to produce a stream by *concatss.*

A binary tree can be defined, as in section 3.7, and the function for flattening the binary tree to a list is:

$$(btree1\; ()\; g\; u) \qquad \textbf{where} \qquad g\; x\; y\; z \;=\; concat(y, x, z).$$

This produces a *list* of nodes of a tree. In order to walk through the nodes of the tree one step at a time, the stream for the binary tree is produced by applying

$$(\textit{btree}\,1 \,()\, g \; \textit{us}) \qquad \textbf{where} \qquad g\; x\; y\; z \;=\; \textit{concatss}(y,\, x,\, z)$$

to the binary tree. Alternative scanning methods can be obtained by permuting x, y, and z in the argument of *concatss*.

Confluent streams.   It is clear that any treelike data structure can be represented by a function like a stream. An infinite binary tree, for example, can be represented by a function which when applied to the null list produces the root and two functions representing the left and right subtrees. A function for generating the trees in which the subtrees depend on the root can be defined by using:

$$\textbf{def rec}\; \textit{genbtree}\, f\, g\, x\, () \;=\; x,\, \textit{genbtree}\, f\, g\, (f\, x),\, \textit{genbtree}\, f\, g\, (g\, x)$$

The binary tree which contains the compositions of n (i.e., all lists of positive integers whose sum is n) on level n is denoted by

$$\textit{genbtree}\, f\, g\, (u\; 1)$$

where

$$f\, x \;=\; ((h\, x) + 1){:}(t\, x) \text{ and } g\, x \;=\; 1{:}x$$

This generates the infinite binary tree whose top is given in Fig. 3.10.

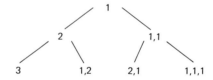

Fig. 3.10   The top levels of an infinite binary tree

The empty binary tree is defined as *emptys* $=$ (*genbtree* $I\, I\, \textit{end}$) and the *empty* predicate as *empty* $f = (\textit{first}(f\,())) = \textit{end}$. The functions *roots*, *lefts*, and *rights* can be defined as the *first*, *second*, and *third* of the result of applying the binary tree to the null list. Functions analogous to those on streams can be constructed for trees. For example, the function *prune*

> **def rec** *prune* $p\; s\; =$
>> **let** $x,\, y,\, z = s()$
>> **if** $p\; x$
>> **then** *emptys*
>> **else** $\lambda().x,\, \textit{prune}\; p\; y,\, \textit{prune}\; p\; z$

converts an infinite tree into a finite one, and is analogous to *untils* for streams. The binary tree for the compositions down to level *n* is obtained by applying (*prune*((*greater n*) • *sum*)).

These infinite tree-processing techniques can be used to solve a problem mentioned by Dijkstra [3–5] and attributed to Weizenbaum. Given an integer *n,* the problem is to write a program to find the smallest number that can be decomposed into the sum of two n^{th} powers in at least two different nontrivial ways. The relevant data structure is an infinite tree of pairs, which starts as follows:

$$(0, 0) - (0, 1) - (0, 2) - (0, 3) - \ldots$$
$$|$$
$$(1, 1) - (1, 2) - (1, 3) - (1, 4) - \ldots$$
$$|$$
$$(2, 2) - (2, 3) - (2, 4) - (2, 5) - \ldots$$
$$|$$
$$:::$$

The structure is an (integer-pair)-tree, where a tree is defined as follows:

- An A-tree

    has a *root* which is an A
    and a *left* which is an A-tree
    and a *right* which is an A-sequence.

The required tree is

> *gen F G* (0, 0)
> > **where** $F(x, y) = x + 1, y + 1$
> > **and** $G(x, y) = x, y + 1$
> > **and rec** *gen f g x* () $= x$, *gen f g*(*f x*), *generate g*(*g x*)

The infinite tree formed by mapping with $H(x, y) = x^n + y^n$ has the property that its root is smaller than any root in its subtrees. The resulting tree can therefore be sorted by taking the root as the first in the sorted sequence, and then merging the two subtrees. The mapping function is

> **def rec** *mapt f x* () $=$ **let** *root, left, right* $= x()$
> > *f root, mapt f left, maps f right*

The tree can now be sorted using the function *sort,* which produces a stream of sorted numbers from the tree.

$$\textbf{def rec } sort\ x\ =\ \textbf{let } root,\ left,\ right\ =\ x()$$
$$\lambda().\ root,\ sort(merge\ left\ right)$$

where rec
$$merge\ x\ y\ =$$
$$\textbf{let } a,\ b,\ c\ =\ x()$$
$$\textbf{let } d,\ e\ =\ y()$$
$$\textbf{if } a < d$$
$$\textbf{then } \lambda().a,\ (merge\ b\ c),\ y$$
$$\textbf{else } \lambda().d,\ x,\ e$$

Next we find the first repeated member of this stream using

$$\textbf{def rec } repeat\ s\ =\ \textbf{let } x,\ y\ =\ s()$$
$$\textbf{let } u,\ v\ =\ y()$$
$$\textbf{if } x\ =\ u$$
$$\textbf{then } x$$
$$\textbf{else } repeat\ y$$

The whole function is

$$\textbf{def } find\ n\ =\ \textbf{let } tree\ =\ gen\ F\ G\ (0,\ 0)$$
$$\textbf{where } F(x,\ y)\ =\ x + 1,\ y + 1$$
$$\textbf{and } G(x,\ y)\ =\ x,\ y + 1$$
$$repeat\,(sort\,(mapt\ H\ tree)))$$
$$\textbf{where } H(x,\ y)\ =\ x^n + y^n$$

Destructive streams.   The stream functions defined above contain no assignment statements, which means that the stream s still exists after it has been applied to the null list. It is often the case that s is no longer needed after it has been applied; and, in such instances, it can be implemented by a destructive routine which contains private storage within itself to record the *tail* of the sequence. When the stream is applied to the null list the private storage is reset by an assignment statement. Such a routine therefore destroys the original stream when it is applied to the null list. An **own** variable feature was introduced in ALGOL 60 to supply this kind of private storage, although there are some questions about how it should be implemented. The intention was to have a variable which was local to a block but which unlike a normal local had a value which survived the activation of the block.

The version of the own variable introduced here is attached to a procedure rather than a block, and, like a stream, depends on using a particular
expression construction which can be used in any context. It does not rely
on the addition of any special mechanical devices for its specification or
implementation. The only rule used is that the body of a lambda expression
is evaluated only when the function is applied. The expression construction
which introduces an **own** variable is a function which is written as an expression whose operator is a lambda expression which in turn has a body
which is a lambda expression. In other words, an expression of the form

$$(\lambda x.\lambda y.\ M)N.$$

The **own** variable in the expression above is $x;$ when the expression is
evaluated x takes as its initial value the value of N. To guard against assignments to this variable from outside, the initial value should be copied,
therefore the expression $(\lambda x.\ \lambda y.M)(copy\ N)$ is better, since it ensures that
the only way of assigning to x is by an assignment statement of the form
$x := E$ within M. It should be noted that the expression above differs from
the construction

$$\lambda y.((\lambda x.M)N)$$

in which the *body* of the lambda expression rather than the whole lambda
expression is qualified. This construction corresponds to a procedure whose
body is a block, and the x is used in a conventional way as a local. Although
the construction $\lambda y.((\lambda x.M)N)$ is possible in ALGOL 60 or PL/I, the construction $(\lambda x.\ \lambda y.M)N$ has no counterpart.

Cascading streams.   Streams, like coroutines, are most useful in specifying
a cascade of editing processes. An example of the method of constructing
a program using streams follows. The problem is taken from Dijkstra [3–6].
The input is made up of a sequence of words composed of letters, separated
by any number of spaces and terminated by spaces and a point, and is
assumed to be a character-stream called *rnc*. The required sequence of
characters replaces the separating spaces by just one space, reverses every
other word, and is terminated with a point.

The function (*while letter*) takes a word from the head of the input and
the function (*while space*) takes spaces until it finds a nonspace. The function
absorbword is defined as:

> **def** *absorbword s* = **let** *w, s*1 = (*while letter s*)
>
>   **let** *sps, s*2 = (*while space s*1)
>
>   *w, s*2

To produce a word-stream this function has to be applied repeatedly by
applying (*next absorbword*) to the character-stream. The point at the end

(in fact any nonletter) will give rise to a tail made up of an infinite sequence of empty lists. Now every other word has to be reversed. The sequence (*generate not* **false**) can be generated to record whether to reverse a word or not. The two streams can then be merged to form a resulting stream using the function

$$(zips\ g)\ \textbf{where}\ g\ x\ y\ =\ \textbf{if}\ y\ \textbf{then}\ reverse\ x\ \textbf{else}\ x$$

which when applied to the word stream (*next absorbword rnc*) and (*generate not* **false**) produces a word-stream with alternate words reversed. The next step is to produce a character stream from a word stream. The spaces and point have also to be inserted. The spaces can be inserted by prefixing a space to every word except the first. This can be done by applying (*maps* (*prefix space*)). Finally the whole character stream is obtained by applying *concatl* and postfixing a point. The whole program becomes:

> **let** $s1$ = *zips g* (*next absorbword rnc*) (*generate not* **false**)
>
> **where** $g\ x\ y$ = **if** y **then** *reverse* x **else** x
>
> **let** $s2$ = *concatl*(*untils*(*null* · *t*)(*maps*(*prefix space*)$s1$))
>
> *postfixs point* (*ts s2*)
>
> **where** *postfixs x y* = *appendl y* (*us x*)

As in all programming systems, there is some choice in the strategy which can be adopted. One of the important decisions seems to be the stage at which one changes from dealing with infinite streams to dealing with streams which represent lists. If a stream is denoted by an expression and is not named, or if it is named and the name is only used once, then it is valid to use a destructive representation of the stream. All the streams in the example above can be replaced by destructive streams.

It should be clear from these examples that any function that operates on, or produces, tree-like data structures can be adapted to operate on, or produce, stream functions.

3.11 COMBINATORIAL CONFIGURATIONS

A set can be represented by a list in a nonunique way, and a list can be represented by a stream. It is often possible to write a program that constructs a set in the most straightforward manner and then to systematically change the constructing operations so that the program constructs a stream. Several examples of programs for generating combinatorial configurations are given in this section.

Permutations.   Certain differential operators are introduced which correspond to programs which reduce a configuration to a simpler form. These operators can be used to obtain the coefficients when one generating function is expanded in terms of others. The differential operators can be given an interpretation by specifying programs for the steps to be carried out in the differentiation. By adding such an interpretation it is possible to treat the differential operator as the program that produces the *set* of configurations itself rather than merely the size of the set.

The simplest example of this technique of elaborating differential operators is to produce a program for generating permutations by interpreting the operations performed when the differential operator $(d/dx)^3$ is applied to x^3. Consider the result of applying d/dx to x^3, written out as xxx. The differential operator can be considered to replace an x by a 1 in three different ways, producing:

$$(d/dx)x^3 = 1.x.x + x.1.x + x.x.1 = 3x^2.$$

These minor operations of changing x's to 1's can be recorded in the three diagrams below:

$$100 \quad 010 \quad 001$$

in which the 1 indicates the position of the x that has been obliterated. If d/dx is now applied to the result then one x is obliterated in all possible ways from all three terms, giving:

$$1.x.x + x.1.x + x.x.1$$
$$1.1.x + 1.x.1 + 1.1.x + x.1.1 + 1.1.x + 1.x.1.$$

The corresponding diagrams are:

$$100 \quad 010 \quad 001$$
$$100 \quad 100 \quad 010 \quad 010 \quad 001 \quad 001$$
$$010 \quad 001 \quad 100 \quad 001 \quad 100 \quad 010.$$

If the differential operator is applied for the third time then the final x is obliterated from each term to give the $(d/dx)^3x^3 = 6$ diagrams or permutations below:

$$100 \quad 100 \quad 010 \quad 010 \quad 001 \quad 001$$
$$010 \quad 001 \quad 100 \quad 001 \quad 100 \quad 010$$
$$001 \quad 010 \quad 001 \quad 100 \quad 010 \quad 100$$

This is a simple example of a method of interpreting a skeletal piece of mathematics by giving it an operator or program interpretation. In this case the number 6, which is the value of $(d/dx)^3x^3$, is the number of diagrams constructed in this way. When interpreted, the operators create a set of configurations and the value of the expression is the number of different configurations formed. It is often the case that the most straightforward way of writing programs which generate combinatorial sets is by considering a tree structure. The method of producing permutations given above is one example. Frequently, however, the combinatorial configurations are required one at a time so that they can be processed by another program. By using the technique of stream functions, discussed in the last section, the program for generating the configurations one at a time can be simply specified.

Combinations. The simplest method of writing a program for producing all combinations of m items chosen from n is to follow the generating function or recurrence relation. Thus

$$(1 + x)^n = (1 + x)(1 + x)^{n-1} = (1 + x)^{n-1} + x(1 + x)^{n-1}$$
$$\binom{n}{m} = \binom{n-1}{m} + \binom{n-1}{m-1}$$

so that
The corresponding program is:

def rec *comb n m* =
    if $m < 0 \lor m > n$
    then ()
    else *append*$(0:comb\ m(n - 1))(1:comb(m - 1)(n - 1))$

This can then be converted to a function which produces a *stream* of combinations as follows:

def rec *comb n m* =
    if $m < 0 \lor m > n$
    then *nullists*
    else *appendl*$(prefixs\ 0(comb\ m(n - 1))$
                      $(prefixs\ 1(comb(m - 1)(n - 1))$

Programs to generate permutations. One program, that is associated with the performance of the differential operator above, first selects the first, second, or third member of the list (1, 2, 3) and removes it to form the last item in a permutation. It next selects the first or second member of the remaining list to form the last but one element. The remaining element is then taken as first element. The tree below follows the operations of the differential operator:

$$\cdots$$

$$\cdot\cdot 1 \quad \cdot\cdot 2 \quad \cdot\cdot 3$$
$$\cdot 21 \quad \cdot 31 \quad \cdot 12 \quad \cdot 32 \quad \cdot 13 \quad \cdot 23$$
$$321 \quad 231 \quad 312 \quad 132 \quad 213 \quad 123$$

The permutation can be uniquely specified by the list of positions used to create it from the list (1, 2, 3). This list of positions is called its signature. The same signature can be used to construct the inverse permutation. This construction relies on the fact that in a permutation $x_1\ x_2\ x_3\ x_4\ \dots\ x_n$ the number of elements less than and to the right of $j = x_i$ is equal to the number of elements greater than and to the left of i in the inverse. To construct the inverse permutation, the signature is read in reverse order and the permutations of n elements are obtained from the permutations of $n - 1$ by inserting n into the position determined by the next signature element. There are n possible positions in $xxxx \dots x$ in which to place a 1. Suppose that the number n is placed in all positions of each permutation of $\{1, 2, 3, 4, \dots, n - 1\}$. The positions into which n can be inserted are numbered 0, 1, 2, 3, 4, $\dots$, n from the end of the permutation. The list of positions into which the numbers 1, 2, 3, 4, $\dots$ are inserted in that order is the signature and is another way to specify the permutation. Fig. 3.11 gives a tree which demonstrates how permutations of $\{1, 2, 3, 4\}$ are generated by this method. The program for producing permutations of 1 to n by this method is:

def rec *perms n* =

    if *n* = 0

    then ()

    else *insert n* (*perms*(*n* − 1))

    where *insert n x* = *concat*(*map*(*put n*)*x*)

        where rec *put n x* =

           if *null x*

           then (*n*:()):()

           else (*n*:*x*):*map*(*prefix*(*h x*))(*put n* (*t x*))

This function can be converted, using the methods given in Section 3.10, to the function that produces the *stream* of permutations, rather than a list. Within each group of four permutations obtained from one permutation of $\{1, 2, 3\}$ in Fig. 3.11 the next can be obtained from the one above by an interchange of adjacent elements. If the immediate subtrees of each root which contains an odd permutation are reversed then it is possible to proceed from one group to the next by the interchange of adjacent elements. Consequently the whole set of permutations can be generated from 1, 2, 3 $\dots$ n by interchanging adjacent elements. A tree for this method of generation is given in Fig. 3.12. This method of generating permutations is known as the 'Johnson-Trotter algorithm' [3–12, 3–15]. One way to write the program is to first generate the signature tree, the first four levels of which are shown in Fig. 3.13.

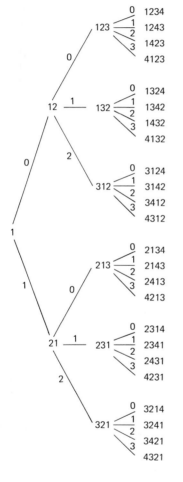

Fig. 3.11 Generating permutations by insertion

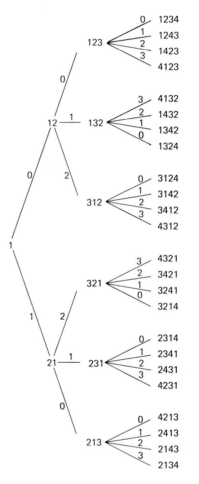

Fig. 3.12 Generating permutations by adjacent transpositions

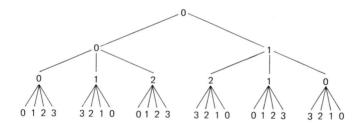

Fig. 3.13 A signature tree

This tree is then scanned using a prefix scan to produce the sequence

$$00001231321020123123210101230 0321$$

the zero's are next removed, because each number is going to represent an interchange or passage from a node to its right neighbor. The resulting string of numbers is:

$$12313212123123211123321$$

The numbers in this sequence now represent the interchanges which have to be made to 1234 to generate the whole sequence of permutations. The number i is to be interpreted as an instruction to interchange the elements in positions $4 - i$ and $5 - i$. The resulting permutations are given in Fig. 3.12, reading from top to bottom. The program can be written as a function which produces the sequence as a stream. Streams for each level of the tree are first provided, namely:

$$\textbf{def } up\ n\ =\ whiles\ (\leq n)(generate(+\ 1)0)$$
$$\textbf{def } down\ n\ =\ whiles\ (\geq 0)(generate(-\ 1)n)$$
$$\textbf{def rec } genalt\ n\ ()\ =\ up\ n,\ (\lambda().\ down\ n,\ genalt\ n)$$

The result of applying *genalt* to n is a stream of streams. The first few examples are:

$$genalt\ 0\ =\ u\ 0,\ u\ 0,\ u\ 0,\ \ldots$$
$$genalt\ 1\ =\ (0,\ 1),\ (1,\ 0),\ (0,\ 1),\ (1,\ 0),\ \ldots$$
$$genalt\ 2\ =\ (0,\ 1,\ 2),\ (2,\ 1,\ 0),\ (0,\ 1,\ 2),\ \ldots$$

These levels of the tree are then put together by using the function *trees*, defined below:

$$\textbf{def } trees\ n\ =$$
$$trees1\ n\ 0$$

$$\textbf{where rec } \textit{trees } n\ x\ =$$
$$\textbf{if } x\ =\ n$$
$$\textbf{then } \textit{generate I nullists}$$
$$\textbf{else } \textit{next } (\textit{first } (x\ +\ 1))$$
$$(\textit{zips ctree } (\textit{concats}(\textit{genalt } x)))$$
$$(\textit{trees}1\ n\ (x\ +\ 1)))$$

$$\textbf{where rec } \textit{first } n\ s\ =$$
$$\textbf{if } n\ =\ 0$$
$$\textbf{then } (),\ s$$
$$\textbf{else let } y,\ z\ =\ \textit{first}\,(n\ -\ 1)(\textit{ts } s)$$
$$\textit{hs } s{:}y,\ z$$

The function *trees n* produces an infinite stream of trees. The first one has to be selected and the prefix scan function is

$$\textbf{def rec } \textit{prescan } x\ =\ \textbf{let } \textit{root, listing }\ =\ x()$$
$$\lambda().\textit{root, concatl}(\textit{mapl prescan listing})$$

The zero's can be removed by applying (*filterl nonzero*).

Partitions. A partition of an integer *n* is a collection of positive integers whose sum is *n*. The integers in the collection are called the parts of the partition. Two different arrangements of the same parts are the same partition and the parts are written in descending order of magnitude. The partition of 10 into the parts 5, 4 and 1 will be written (5 4 1). An array of nodes called a Ferrer's graph is often used to denote a partition. The number of nodes in the rows are the parts of the partition, and a larger part is placed in a higher row. The Ferrer's graph for the partition (5 4 1) is given below.

```
· · · · ·
 · · · ·
    ·
```

The partitions of the first five numbers are:

1. (1)
2. (2), (1 1)
3. (3), (2 1), (1 1 1)
4. (4), (3 1), (2 2), (2 1 1), (1 1 1 1)
5. (5), (4 1), (3 2), (3 1 1), (2, 2, 1), (2 1 1 1), (1 1 1 1 1)

When parts are repeated an exponent will be used and the last line could be written as:

$$(5), (4\ 1), (3\ 2), (3\ 1^2), (2^2\ 1), (2\ 1^3), (1^5)$$

The partitions can be enumerated by first choosing the number of ones $((1/1 - x))$, then choosing the number of twos $((1/1 - x^2))$, then the number of threes, etc.

$$\text{partitions}(x) = (1/(1-x))(1/(1-x^2))(1/(1-x^3)) \ldots$$
$$= 1 + x + 2x^2 + 3x^3 + 5x^4 + 7x^5 + \ldots$$

Restrictions which are placed upon the number of parts and the size of the parts can be reflected in the restrictions on the size of this product, and upon the nature of its terms. For example the generating function for partitions containing no part greater than k is

$$(1/(1-x))(1/(1-x^2))(1/1 - x^3)) \ldots (1/(1 - x^k))$$

The generating function for partitions into exactly i parts is the coefficient of z^i in:

$$1/(1 - zx)(1 - zx^2)(1 - zx^3) \ldots$$

which has the expansion

$$1 + zx/(1-x) + z^2x^2/(1-x)(1-x^2) + \ldots$$
$$+ z^i x^i/(1-x)(1-x^2) \ldots (1 - x^i)$$

showing that the generating function for partitions into exactly i parts is

$$x^i(1/(1-x))(1/(1-x^2))(1/(1-x^3)) \ldots (1/(1-x^i))$$

This will be written in the abbreviated form $x^i/[i]!$

A program for generating the partition of a number n into exactly i parts can be derived as follows: Since

$$x^i/[i]! = (1 - x^i + x^i)x^i \times [i]!$$
$$= x.x^{i-1}/[i-1]! + x^i.x^i/[i]!,$$

then if $P(n, i)$ is the number of partitions of n into i parts

$$P(n, i) = P(n - 1, i - 1) + P(n - i, i)$$
$$P(n, 0) = 1 \ if \ n = 0$$
$$= 0 \ if \ n > 0$$
$$P(n, i) = 0 \ if \ n < i.$$

If the partitions are represented by lists of integers in ascending order then the following function constructs partitions from the partitions into those having one as their smallest part and those which do not. The program below constructs a list of partitions, but can be adapted to produce a stream.

```
def rec partitions n i =
    if n < i
    then ()
    else if j = 0
        then if n > 0
            then ()
            else ():()
        else append(1:(partitions(n − 1, i − 1))
                    (map(+ 1)(partitions(n − i, i)))
```

REFERENCES AND BIBLIOGRAPHY

Good surveys of data structures and methods of representing them may be found in Berztiss [3–1], Hoare [3–8], D'Imperio [3–10] and Knuth [3–11], Chapter 2. The methods for defining new data structures were introduced into programming by McCarthy, and some programming languages that have this capability are ALGOL 68 [3–16], PASCAL [3–19], and POP2 [3–2]. Streams are very similar to coroutines (Conway, 1963), and similar streaming techniques are discussed in references 3-2, 3-5, 3-6, 3-7, and 3-14.

3–1. Berztiss, A. T., *Data Structures, Theory and Practice,* London and New York, Academic Press, 1971.

3–2. Burstall, R. M., J. S. Collins, and R. J. Popplestone, *Programming in POP-2,* Edinburgh: Edinburgh University Press, 1971.

3–3. Burge, W. H., "Combinatory programming and combinatorial analysis," *IBM J. Res. and Dev.,* Vol. 16, No. 5, 1972, pp. 450–461.

3–4. Conway, M. E., "Design of a separable transition-diagram compiler," *CACM,* Vol. 6, 1963, pp. 396–408.

3–5. Dahl, O. J., and K. Nygaars, "Simula—an ALGOL-based simulation language," *CACM,* Vol. 9, 1966, pp. 671–678.

3–6. Dijkstra, E. W., O. J. Dahl, and C. A. R. Hoare, *Structured Programming,* London and New York: Academic Press, 1972.

3–7. Golomb, S. W., and L. D. Baumert, "Backtrack programming," *JACM,* Vol. 12, 1965, pp. 516–524.

3–8. Hoare, C. A. R., *Record Handling in Programming Languages,* (ed.) F. Genuys, London: Academic Press, 1968, pp. 291–347.

3-9. Holt, A. W., "A mathematical and applied investigation of tree structures," *Ph.D. Thesis,* University of Pennsylvania, 1963.

3-10. D'Imperio, M. E., "Data structures and their representation in storage," *Ann. Rev. Automatic Programming,* Vol. 5, 1969, pp. 1–75.

3-11. Knuth, D. E., *The Art of Computer Programming: Vol. 1: Fundamental Algorithms,* Reading, Mass.: Addison-Wesley, 1968.

3-12. Johnson, S. M., "An algorithm for generating permutations," *Math. Comp.,* Vol. 17, 1963, p. 28.

3-13. Naur, P., "Programming by action clusters," *BIT,* Vol. 9, No. 3, 1969, pp. 250–258.

3-14. Stoy, J. E. and C. Strachey, "OS6—An experimental operating system for a small computer," *Computer J.,* Vol. 15, 1972, No. 2, pp. 117–124; No. 3, pp. 195–203.

3-15. Trotter, H. F., Algorithm 115: Perm, *CACM,* Vol. 5, 1962, pp. 434–435.

3-16. van Wijngaarden, A., "Report on the algorithmic language, ALGOL 68," *Num. Math.,* Vol. 14, 1969, pp. 79–218.

3-17. Wells, M. B., *Elements of Combinatorial Computing,* Oxford: Pergammon Press, 1971.

3-18. Wirth, N., *Systematic Programming,* Englewood Cliffs, N.J.: Prentice-Hall, 1973.

3-19. Wirth, N., "The programming language PASCAL," *Acta Informatica,* Vol. 1, No. 1, 1971, pp. 35–63.

4
Parsing

4.1 INTRODUCTION

This chapter contains several approaches to the problem of recognizing and analyzing the structure of lists of characters. Characters are treated as atomic, recognizable elements and a character-list is usually, following common practice, called a character *string*. A set of character strings constitutes a language, and a *recognizing function* for a language determines whether a given character string belongs to the set. A *parsing function* for a language not only determines whether a string belongs to the language but also produces some object from it.

There are two important classes of languages called *regular* languages and *context-free* languages. These types of sets can be systematically specified and recognizers and parsers systematically produced from the descriptions of the sets. Whereas the parser that naturally corresponds to a regular language can be made deterministic, and the corresponding program derived in a relatively straightforward way, the parser that naturally corresponds to the grammar of a context-free language is nondeterministic, and a problem arises in constructing an efficient parsing program for the language. The techniques in this chapter are more concerned with combining recognizing programs than with the nature of the basic recognizers that are being combined. Analogous techniques can be applied to other types of recognition problem in which a nondeterministic program has to be converted to a deterministic one. The most common parsing program is one that imposes a structural description on the string being recognized. This is a tree which analyzes the phrase structure of the string. Three main parsing strategies are described: top-down, left-corner bottom-up, and right-corner bottom-up. These are descriptions of the way in which this tree is constructed.

4.2   FINITE AUTOMATA AND REGULAR EXPRESSIONS

The simplest type of language is a regular language. A regular language can be recognized using a program based upon a directed graph called a state diagram. The nodes of this graph represent states of the machine, and each edge is labeled with a character. An edge represents the state transformation that is to be made when the character at that edge is found at the head of the string to be recognized. One state is chosen as the *initial* state, and one or more states are chosen as *final* states. A string is recognized if it transforms the initial state into one of the final states. It fails to be recognized if it transforms the initial state into a nonfinal state. In other words, the language recognized is made up of all the paths from an initial to a final state.

An example of a state diagram is given in Fig. 4.1. It recognizes the numbers of ALGOL 60. The nodes are labeled: L_1 is the initial node, and the final nodes are indicated in the diagram by double circles. Note that an abbreviation has been made in Fig. 4.1. The letter d stands for the set of digits 0-9. In general, if all characters of a set transform one state to the same state, it is permissible to use such an abbreviation. Such a diagram is a complete specification of the numbers of ALGOL 60. The answer to the question "Is 537. an ALGOL 60 number?" for example, can be found by passing from node L_1 to L_4. Since L_4 is a nonfinal state, the answer is no.

The graph can also be specified as a set of productions. Each part of the graph of the form

$$A \rightarrow B$$
$$a$$

gives rise to a production which is written as follows:

$$A \rightarrow aB$$

Also if A is a final state, a production

$$A \rightarrow e$$

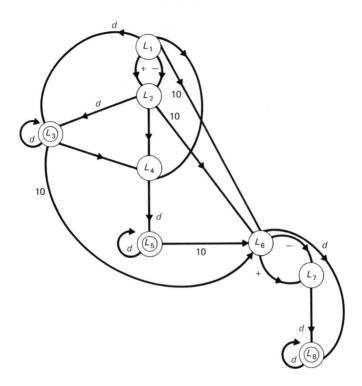

Fig. 4.1 A state diagram for Algol 60 numbers

is added, where *e* stands for the null string. Each node can be considered to describe a language made up of all the paths from that node to a final state. Both the productions and the diagram specify these paths. The production

$$A \rightarrow aB$$

means that one set of paths from *A* to a final state is constructed by first taking path *a* and then taking the set of paths from *B* to a final state. The production

$$A \rightarrow e$$

means that *A* is already a final state; and so the empty string will take it to itself, a final state. The productions that correspond to the state diagram

?

follow. The only information missing is the designation that the initial state is L_1.

$$L_1 \to dL_3 \qquad L_4 \to dL_5$$
$$L_1 \to + L_2 \qquad L_5 \to dL_5$$
$$L_1 \to - L_2 \qquad L_5 \to {}_{10}L_6$$
$$L_1 \to {}_{10}L_6 \qquad L_6 \to + L_7$$
$$L_1 \to .L_4 \qquad L_6 \to - L_7$$
$$L_2 \to .L_4 \qquad L_6 \to dL_8$$
$$L_2 \to {}_{10}L_6 \qquad L_7 \to dL_8$$
$$L_2 \to dL_3 \qquad L_8 \to dL_8$$
$$L_3 \to dL_3 \qquad L_3 \to e$$
$$L_3 \to {}_{10}L_6 \qquad L_5 \to e$$
$$L_3 \to .L_4 \qquad L_8 \to e$$

The program for recognizing strings specified by the productions can be written down immediately from them. For example the recognizing function corresponding to the productions

$$L_6 \to + L_7$$
$$L_6 \to - L_7$$
$$L_6 \to dL_8$$

can be written as follows:

$L_6(s) = $ **if** *null s*

      then false

      else if $h\,s = $ '$+$'

          then $L_7(t\,s)$

          else if $h\,s = $ '$-$'

               then $L_7(t\,s)$

               else if *digit* $(h\,s)$

                  then $L_8(t\,s)$

                  else false

The productions having common left sides are grouped together to produce a function. The example given above corresponds to a nonfinal state. The piece of program that corresponds to a final state, however, must start:

if *null s* **then true**

The functions formed in this way are mutually recursive. However, since their application only occurs in operator position they can be rewritten in program form using assignment statements and **go to**'s in the following program. It is assumed that the program continues at the place labeled *fail,* if the string is not recognized, and at the label called *succeed,* if it is.

$$L_1: \textbf{if } null \ s$$
$$\textbf{then go to } fail$$
$$\textbf{else } x: = h \ s$$
$$s: = t \ x$$
$$\textbf{go to if } x = \text{`+'}$$
$$\textbf{then } L_7$$
$$\textbf{else if } x = \text{`\_'}$$
$$\textbf{then } L_7$$
$$\textbf{else if } digit \ x$$
$$\textbf{then } L_8$$
$$\textbf{else } fail$$

The piece of program which has a label that corresponds to a final state must start:

$$\textbf{if } null \ s$$
$$\textbf{then go to } succeed$$

The state diagram can be considered in two ways: 1) as a method of recognizing a string of characters, and 2) as a method for generating all the strings of the language. In the second case the language corresponding to a node is made up of all paths from the node to a final state.

The productions can also be considered to describe a language or set of strings by a set of equations. This leads to another way to describe languages and to construct their associated parsers. The productions can be considered as a set of equations having two operations between languages called *union* and *Cartesian concatenation,* respectively. The union is set union and the union of two sets A and B is written $A \mid B$ and is defined as follows:

$$A \mid B = \{x \mid x \, \varepsilon \, A \text{ or } x \, \varepsilon \, B\}$$

The Cartesian concatenation of two sets of strings A and B is the set whose members are made up of the concatenation of a member of A followed by a member of B.

$$A \cdot B = \{append \ x \ y \mid x \, \varepsilon \, A, y \, \varepsilon \, B\}$$

Sometimes $+$ is used instead of $|$, and $\cdot$ is often omitted. A third operation called *closure* or *Kleene closure* is defined as follows:

$$A^* = e\,|\,A\,|\,A^2\,|\,A^3\,|\,\ldots\,\textit{where } A^n = A^{n-1} \cdot A$$

The regular languages are described by regular expressions. The simple regular expressions are: 1) a symbol for the empty set, ϕ, 2) a symbol for the set containing one element which is the nullist, e, and 3) the symbols of the alphabet (each a in the alphabet denotes the set $\{a\}$). The compound regular expressions are $A\,|\,B$, $A \cdot B$ and A^*, where A and B are regular expressions. The following identities follow from the definitions. The un-fixed $\cdot$ operator will be omitted in this section.

$$A\,|\,A = A$$
$$A\,|\,B = B\,|\,A$$
$$A\,|\,(B\,|\,C) = (A\,|\,B)\,|\,C$$
$$A(BC) = (AB)C$$
$$AB\,|\,AC = A(B\,|\,C)$$
$$AC\,|\,BC = (A\,|\,B)C$$
$$(A\,|\,B)^* = (A^*\,|\,B^*)^* = (A^*B^*)^*$$
$$Ae = eA = A$$
$$A\phi = \phi A = \phi$$
$$\phi^* = e$$
$$A\,|\,\phi = A$$
$$A^* = e\,|\,AA^*$$

It can also be shown that if $A = BA\,|\,C$ then

$$A = B^*C.$$

This last fact can be used to solve the equations that correspond to the productions. The equations for ALGOL 60 numbers are:

$$L_1 = dL_3\,|\, + L_2\,|\, - L_2\,|\,_{10}L_6\,|\, . \, L_4$$
$$L_2 = . \, L_4\,|\,_{10}L_6\,|\,dL_3$$
$$L_3 = dL_3\,|\,_{10}L_6\,|\, . \, L_4\,|\,e$$
$$L_4 = dL_3$$
$$L_5 = dL_5\,|\,_{10}L_6\,|\,e$$
$$L_6 = + L_7\,|\, - L_7\,|\,dL_8$$
$$L_7 = dL_8$$
$$L_8 = dL_8\,|\,e$$

It follows that, by using the identities above

$$L_8 = d*$$
$$L_7 = dd*$$
$$L_6 = (+\,|\,-)dd*\,|\,dd* = (+\,|\,-\,|\,e)dd*$$
$$L_5 = d*(_{10}L_6\,|\,e)$$
$$L_4 = dd*(_{10}L_6\,|\,e)$$
$$L_3 = d*(_{10}L_6\,|\,.\,L_4\,|\,e)$$
$$L_2 = d\,.\,d*(_{10}L_6\,|\,.\,L_4)\,|\,.\,L_4\,|\,_{10}L_6$$
$$L_1 = (+\,|\,-\,|\,e)L_3$$
$$L_2 = dd*(.\,dd*(_{10}L_6\,|\,e)\,|\,(_{10}L_6\,|\,e))\,|\,.\,dd*(_{10}L_6)\,|\,e)\,|\,_{10}L_6$$
$$\quad = (dd*\,|\,dd*\,.\,dd*\,|\,.\,dd*)(_{10}L_6)\,|\,_{10}L_6$$
$$L_1 = (+\,|\,-\,|\,e)((dd*\,|\,e)\,.\,dd*\,|\,dd*)(_{10}(+\,|\,-\,|\,e)dd*\,|\,e)\,|\,(+\,|\,-\,|\,e)dd*$$

This expression can be rewritten by extracting common subexpressions and using auxiliary definitions as follows:

> *signed* (*number*)
>> **where** *integer* = *dd**
>>> **and** *signed x* = (+ | − | *e*)*x*
>>> **and** *fraction* = . *integer*
>>> **and** *exponent* = $_{10}$*signed*(*integer*)
>>> **and** *decimal* = *integer* | *fraction* | *integer fraction*
>>> **and** *number* = *decimal* | *exponent* | *decimal exponent*

Another type of state diagram can be constructed in which the edges are labeled with regular expressions rather than characters. The stages in the solution can be regarded as transformations of the state diagram in which several paths between nodes are replaced by one path labeled with a regular expression which represents the original set. The regular expression for L_6 can be obtained in this way, as shown in Fig. 4.2. A new final state is introduced, and the diagram is rearranged so that all edges are directed to it.

The reverse process, i.e., constructing a set of productions from a regular expression, is also possible. The productions for regular languages can be put into one of the forms:

$$A \rightarrow e$$
$$A \rightarrow a$$
$$A \rightarrow aB$$

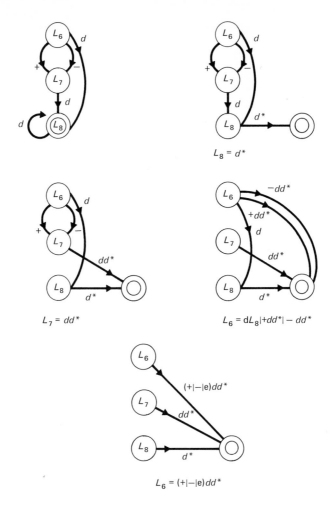

Fig. 4.2   Transforming a state diagram

In the last production the first symbol on the right-hand side is a character and the right-hand side has no more than two symbols. This is called a left-linear grammar. Productions without these restrictions will be considered in the next section.

4.3   CONTEXT-FREE LANGUAGES

Whereas a regular language can be regarded as all the paths through a graph or program graph, a context-free language can be regarded as all the paths through a program graph that has procedures. As with the regular languages, the set of strings of a context-free language can be defined in terms of certain existing sets using the operations of union and Cartesian concatenation. The productions or equations are no longer subject to the restricted left-linear form of those for regular languages, and the right-hand side of an equation can be a finite union of finite Cartesian concatenations. It follows that the definitions can be *self-embedding*, which means that they have an occurrence of a defined symbol in the definitions that is not at the end of a concatenation. The context-free languages include both regular languages and certain languages which have nested phrases and are context-free but not regular.

The symbols are divided into terminal and nonterminal symbols. The nonterminal symbols are the defined symbols, and so occur on the left side of productions; the terminal symbols are used but not defined. Lower case letters will usually be written for terminal symbols, and upper case for nonterminals. The productions can be considered to generate a set of strings by imposing a relation called directly generates (written $\Rightarrow$,) between two strings of symbols. Thus, $x \Rightarrow y$ means that y can be obtained from x by substituting the right-hand side of a production for one occurrence of its left-hand symbol in x. The language defined by a symbol S is the set of strings of terminal symbols related to the string containing the one character S by the reflexive transitive closure of the relation $\Rightarrow$ (written $\Rightarrow^*$).

The *transitive closure* of a relation can be defined as follows. The product of two relations r and s is the relation that exists between x and y if and only if there is a u such that x bears the relation r to u and u bears the relation s to y. In other words if r and s are treated as functions of type $(A \rightarrow A\text{-list})$ or $(A \rightarrow A\text{-stream})$, then their product can be defined as follows:

$$\textbf{def } prodf\ r\ s\ x\ =\ sumset\ (map\ s(r\ x))$$

$$\textbf{where } sumset\ =\ list1\ ()\ union\ I$$

A relation can be expressed as a directed graph in which the nodes represent the objects, and the edges the relation. There is a directed edge from x to y if and only if y is related to x. The nth power of a relation can be expressed as:

$$\textbf{def rec } power\ n\ r\ =$$

$$\textbf{if } n\ =\ 0$$

$$\textbf{then } I$$

$$\textbf{else } prodf\ r(power(n\ -\ 1)r)$$

The relation (*power n r*) relates node x in the graph to all those nodes that can be reached by following a directed path with n edges. The relation (*power n r*) can also be written r^n. The transitive closure of a relation relates a node of the graph to all the nodes that can be reached by following a directed path. In other words

$$r^+ = \bigcup_{n=1}^{\infty} r^n$$

The reflexive transitive closure includes the identity relation and is defined below:

$$r^* = \bigcup_{n=0}^{\infty} r^n$$

Several useful relations can be derived from the productions of a context-free language. Each nonterminal symbol X can be related, by a relation called *left*, to the symbols that occur as leftmost symbols of the right-hand-side productions for X. The graph of the relation *left* for the productions below is given in Fig. 4.3.

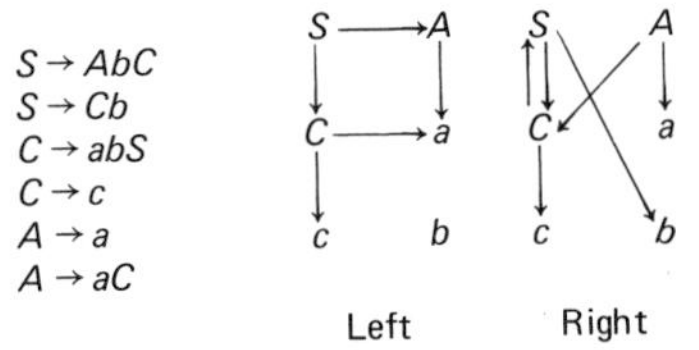

$$
\begin{aligned}
S &\to AbC \\
S &\to Cb \\
C &\to abS \\
C &\to c \\
A &\to a \\
A &\to aC
\end{aligned}
$$

Left   Right

Fig. 4.3 The graphs of the relations left and right

The transitive closure of *left*, i.e., $left^+$, is the relation between the symbol X and all the symbols that could occur at the head of the strings derived from X. In the case of the productions in Fig. 4.3 the relation *left* is:

$$
\begin{aligned}
left^+ \ `S` &= \{A, C, a, c\} \\
left^+ \ `A` &= \{a\} \\
left^+ \ `C` &= \{c\}
\end{aligned}
$$

If $left^+ \ X$ includes X then X is said to be left recursive. The relation *first* is defined to be the restriction of $left^+$ to terminal symbols, i.e.

$$first \ X = \{y \ \varepsilon \ left^+ \ X \ | \ y \ \varepsilon \ terminal\}$$

In a similar way, *right* is the relation between a symbol X and those symbols occurring at the extreme right of derivations from X. If $right^+ \ X$ includes X, then X is said to be right recursive. From Fig. 4.3 it can be seen that S is right recursive. The relation *last* is defined to be the restriction of $right^+$

to terminal symbols, thus

$$last\ X = \{y\ \varepsilon\ right^+\ X \mid y\ \varepsilon\ terminal\}$$

There are two systematic ways to generate a string of a language: 1) to always substitute for the leftmost nonterminal in the string, and 2) to always substitute for the rightmost. The corresponding relations will be written $\Rightarrow_L$ and $\Rightarrow_R$. Consider the language defined by S and the productions in Fig. 4.4, for example. The tree in Fig. 4.4 is a demonstration that the string *aabcbbc* belongs to the language.

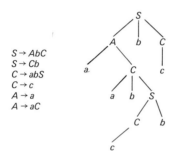

$$
\begin{aligned}
&S \rightarrow AbC \\
&S \rightarrow Cb \\
&C \rightarrow abS \\
&C \rightarrow c \\
&A \rightarrow a \\
&A \rightarrow aC
\end{aligned}
$$

Fig. 4.4   The structural description of a string

This tree might have been generated by leftmost derivations as in (a), or using rightmost as in (b), or by some other method (c).

(a) $S \Rightarrow_L AbC \Rightarrow_L aCbC \Rightarrow_L aabSbC$
    $\Rightarrow_L aabCbbC \Rightarrow_L aabcbbC \Rightarrow_L aabcbbc$

(b) $S \Rightarrow_R AbC \Rightarrow_R Abc \Rightarrow_R aCbc \Rightarrow_R aabSbc$
    $\Rightarrow_R aabCbbc \Rightarrow_R aabcbbc$

(c) $S \Rightarrow_R AbC \Rightarrow_L aCbC \Rightarrow_R aabSbC$
    $\Rightarrow_R aabSbc \Rightarrow_R aabCbbc \Rightarrow_R aabcbbc$

The tree is called the structural description of the string. Its end points are labeled with terminal symbols, and nonterminal nodes are labeled with nonterminal symbols. The fragment of the tree corresponding to a use of the production $X \rightarrow ABCD$

is written

which is a tree whose root is 'X' and whose listing contains A, B, C, D.

The problem in writing parsing programs for context-free languages is to produce the inverse operation; that is, given a string and the productions, to impose a structural description upon the string. The parsing programs have additional actions that usually come into play after the point where an instance of the right-hand side of a production has been recognized. The left-hand side replaces the instance of the right-hand side, and this replacement is called a reduction.

The same set of strings can be described by different sets of productions. It is often useful to be able to rearrange the productions to produce a more restricted set of productions that describe the same set. For example it is often useful to remove symbols which can produce the empty string.

It is possible to determine whether a nonterminal symbol can produce the empty string by a marking process. First mark the empty-string symbols. Then find an unmarked left side whose right-hand side contains only marked symbols. Mark this left-hand symbol and all its occurrences in the right sides of productions, and repeat the process until no further marking is possible. A nonterminal can produce the null string only if it is marked.

The same procedure can be used to determine whether or not a nonterminal symbol can produce any terminal strings. In this case, first mark all terminal symbols, and then carry out the marking process as before. The unmarked symbols at the end of this algorithm are those that produce no terminal strings. Such nonterminals are useless and can be removed, together with any production that includes them in the right-hand side, with no effect on the language.

4.4   EXPRESSIONS THAT DESCRIBE LANGUAGES

Productions that describe a language can be recast as definitions of new sets in terms of old sets using the functional notation introduced in Chapter 1. It will be seen that the description of the set can be reinterpreted as a description of the recognizer for members of the set and then elaborated further to provide parsers for the language.

In the production of a context-free language, a terminal symbol is used to describe the set with one member which is a 1-list having the symbol as its sole element. A function called Q will be introduced in order to be able to distinguish between the names of sets and names of strings. The function Q operates on an individual and produces the set of individuals that are equal to it. The expression $Q`a`$ therefore denotes the set whose members are all equal to a unit list whose sole element is the character a.

The expressions that describe sets of strings of characters are constructed using the operators for set union ($|$) and concatenated cross product ($\cdot$). They both operate on two character-string sets and produce a character-string set.

The empty set is called *empty* (ϕ); the set whose sole member is the nullist is called *nullistset* (e); and the set of unit lists whose items are the characters of the alphabet currently being used is called *anycharacter*. The Cartesian concatenation operator has been made explicit so that juxtaposition is now reserved for the application of a prefixed operator such as Q. One consequence of this change of notation is that parentheses can be used without confusion with their quoted versions for merely grouping the expressions that denote the set. Another consequence of treating the productions as definitions of sets is the possibility of defining new functions from sets to sets, such as

$$\textbf{def rec } list\ a\ b\ =\ b\ |\ b\cdot a\cdot list\ a\ b$$

The set (*list a b*) is the set of nonnull lists of *b*'s separated by *a*'s. The star function, which occurs in regular expressions, can be defined as follows:

$$\textbf{def rec } star\ x\ =\ x\cdot star\ x\ |\ nullistset$$

The set (*star x*) consists of strings formed by any number of repetitions of the phrase *x*. The following functions are useful for specifying repeated phrases.

$$\textbf{def } perhaps\ x\ =\ x\ |\ nullistset$$
$$\textbf{def rec } upto\ n\ x\ =$$
$$\quad \textbf{if } n\ =\ 0$$
$$\quad \textbf{then } nullistset$$
$$\quad \textbf{else } x\cdot upto\ (n-1)\ x\ |\ nullistset$$
$$\textbf{def rec } exactly\ n\ x\ =$$
$$\quad \textbf{if } n\ =\ 0$$
$$\quad \textbf{then } nullistset$$
$$\quad \textbf{else } x\cdot exactly\ (n-1)\ x$$
$$\textbf{def } atleast\ n\ x\ =\ exactly\ n\ x\cdot star\ x$$
$$\textbf{def rec } qualify\ x\ y\ =\ x\ |\ qualify\ x\ y\cdot y$$

The functions *perhaps, upto, exactly* and *atleast* operate on an integer *n* and a phrase *x* to produce the set of phrases containing *m* occurrences of *x*, where $m = 0$ or 1, $0 \le m \le n$, $m = n$, and $m \ge n$, respectively.

The concatenated cross product and union operations can be extended to operate on a list of sets as follows:

def rec *cclist x* =
　　if *null x*
　　then *nullistset*
　　else *h x* • *cclist* (*t x*)
or **def** *cclist* = *list*1 *nullistset cc I*
　　　　　where *cc x y* = *x* • *y*
def rec *unionlist x* =
　　if *null x*
　　then *empty*
　　else *h x* | *unionlist* (*t x*)
or **def** *unionlist* = *list*1 *empty un I*
　　where *un x y* = *x* | *y*

From abstract to concrete syntax.   Although the abstract syntax specifies the structure independently of any written or other representation, there are frequently one or more written formats that are naturally associated with each structure. The description of a written representation (or its concrete syntax) can also be written down from the structure description by specifying an ordering of the components and adding extra characters to avoid ambiguity. Suppose a string of characters has been chosen to represent each atomic component of a structure. Then the set of strings that correspond to a set of structures can be described by an expression which has the same structure as the description of the abstract syntax. The language that corresponds to a direct union is a union of languages. The language associated with

$$du(A, B, C, D, E)$$

is

$$A \mid B \mid C \mid D \mid E$$

where A, B, etc. are now the languages for A, B, etc. In a similar way the language for

$$cp(A, B, C, D, E)$$

is

$$A \cdot B \cdot C \cdot D \cdot E$$

Lists.   The definition of an A-list follows:

$$\textbf{def } list\ A\ =\ du(cp\,(),\ cp\,(A,\ list\ A\,))$$

A language for lists is given by either

$$list\ x\ =\ empty\,|\,x\boldsymbol{\cdot} list\ x\ \text{ or }\ =\ empty\,|\,(list\ x)\boldsymbol{\cdot}x$$

which is the language composed of a phrase x repeated any number of times. If the list items are to be separated then an appropriate function is

$$\textbf{def } lists\ y\ x\ =\ empty\,|\,nlists\ y\ x$$
$$\textbf{where rec } nlists\ y\ x\ =\ x\,|\,x\boldsymbol{\cdot}y\boldsymbol{\cdot}nlists\ y\ x$$

which has the additional complication of the necessity to avoid strings which end with a separator y.

List structures.   The abstract syntax of list structures is:

$$\textbf{def rec } liststructure\ A\ =\ du(A,\ list\ (liststructure\ A\,))$$

One language for list structures is formed by bracketing the nonatomic list structures

$$\textbf{def } ls\ z\ y\ x\ =\ x\,|\,z\boldsymbol{\cdot}list\,(ls\ x)\boldsymbol{\cdot}y$$
$$\text{or }\textbf{def } lss\ s\ z\ y\ x\ =\ x\,|\,z\boldsymbol{\cdot}lists\ s\,(lss\ s\ z\ y\ x)\boldsymbol{\cdot}y$$

if the list items are to be separated. It follows that (*lss comma open close x*) where

$$\textbf{def } open\ =\ Q\text{`}(\text{'}$$
$$\textbf{def } close\ =\ Q\text{`})\text{'}$$
$$\textbf{def } comma\ =\ Q\text{`},\text{'}$$

is a language for list structures.

Trees and forests.   The structure of trees and forests is given by

$$\textbf{def rec } tree\ A\ =\ cp\,(A,\ forest\ A\,)$$
$$\textbf{and } forest\ A\ =\ list\,(tree\ A\,)$$

There are several languages for trees and forests; perhaps the simplest is

$$\textbf{def rec } tree\ x\ =\ open\boldsymbol{\cdot}x\boldsymbol{\cdot}forest\ x\boldsymbol{\cdot}close$$
$$\textbf{and } forest\ x\ =\ list\,(tree\ x)$$

in which the trees are bracketed, the forests are listed, and the root is the first item in the list. An example is:

$$(f,\ x,\ (g,\ y,\ z)).$$

The set (*tree empty*) is the language of *lattice permutations.* Another language for trees is that of a functional notation having the operations of application indicated by juxtaposition of root and listing and by arguments within brackets. For example:

def rec *tree x* = *x • forest x*

and *forest x* = *open • lists comma (tree x) • close*

An example is:

$$f(x, g(y, z))$$

4.5   TOP-DOWN PARSING

A pushdown automaton.   The top-down parsing program that naturally corresponds to a context-free language is nondeterministic and can be expressed as a pushdown automaton (pda). The state of the pda has two lists: 1) the string of terminal symbols which is being recognized, and 2) a pushdown list which can contain both terminal and nonterminal symbols. The state transformations which can take place will be written using the notation

$$(p, q) \rightarrow (r, s)$$

where p and r are strings of symbols, and q and s are strings of nonterminal symbols. The heads of p and r are at the right, and the heads of q and s are at the left. Each state transformation is conditional. If p is an initial segment of the pushdown list, and q is an initial segment of the input string, then they are both to be removed and replaced by the strings r and s respectively. The state transitions for a top-down pushdown automaton can be written down directly from the productions. Each production of the language of the form

$$A \rightarrow U_1 U_2 U_3 \ldots U_n$$

corresponds to the following state transition of the top-down pda.

$$(A, ()) \rightarrow (U_n U_{n-1} \ldots U_3 U_2 U_1, ()),$$

and each terminal symbol a corresponds to the state transition

$$(a, a) \rightarrow ((), ()).$$

In other words, when a nonterminal symbol is found at the head of the pushdown, it is replaced by the right-hand side of its production. The input string is unchanged. When a terminal symbol is found both at the head

of the pushdown and at the head of the string, then both are removed. If:
1) the top down pda is started in the state

$$(S, x)$$

where x is the string to be recognized and S is the symbol for the language
to be recognized, and 2) a series of state transitions can be found that lead
to the state

$$((), ())$$

then the string has been recognized as a member of the language S. The
top down pda that corresponds to the productions given above is given in
Fig. 4.5. The transitions are numbered and Fig. 4.6 gives the sequence of
states leading to the recognition of the string *aabcbbc* as a member of language S.

Productions	State transitions	
$S \rightarrow AbC$	$(S, ()) \rightarrow (CbA, ())$	(1)
$S \rightarrow Cb$	$(S, ()) \rightarrow (bC, ())$	(2)
$C \rightarrow abS$	$(C, ()) \rightarrow (Sba, ())$	(3)
$C \rightarrow c$	$(C, ()) \rightarrow (c, ())$	(4)
$A \rightarrow a$	$(A, ()) \rightarrow (a, ())$	(5)
$A \rightarrow aC$	$(A, ()) \rightarrow (Ca, ())$	(6)
	$(a, a) \rightarrow ((), ())$	(7)
	$(b, b) \rightarrow ((), ())$	(8)
	$(c, c) \rightarrow ((), ())$	(9)

Fig. 4.5　The correspondence between productions and state transitions

Pushdown list, head at right	Input string, head at left	Transformation number
S	*aabcbbc*	1
CbA	*aabcbbc*	6
$CbCa$	*aabcbbc*	7
CbC	*abcbbc*	3
$CbSba$	*abcbbc*	7
$CbSb$	*bcbbc*	8
CbS	*cbbc*	2
$CbbC$	*cbbc*	4
$Cbbc$	*cbbc*	9
Cbb	*bbc*	8
Cb	*bc*	8
C	*c*	4
c	*c*	9
$()$	$()$	

Fig. 4.6　The sequence of states that recognize a string

Since the nonterminal symbol at the head of the pushdown is expanded before any symbol beneath it, the recognition proceeds by expanding left-most derivations and tests whether the strings derived in this way match the input string. There may be no sequence of transitions which leads to a final state. In this case the string does not belong to the language. If there is one sequence then the string can be recognized and a structural description can be imposed. It can be the case that more than one sequence of transitions lead to a final state. When this happens the grammar is said to ambiguous. Although there is no general test for ambiguous grammars it is often possible to test whether a grammar is of a type which is known to be unambiguous.

Parsing relations.   The same pushdown automaton can be described from a different point of view in terms of parsing relations. Suppose that for each symbol, whether terminal or nonterminal, there is a relation from a string to set of strings. The relation that corresponds to a particular phrase is between one string and another from which an initial segment which is an instance of that phrase has been removed. The relation can be considered a function from a string to a set of strings. If the function does not find its phrase at the head of the string, the result is the empty set. The relation that corresponds to each terminal symbol tests whether that symbol is at the head of the string. If so, then the result is a set containing one member, the tail of the string. If not, then the result is the empty set. If sets are represented by stream functions then the function Q, defined below, operates on a terminal symbol to produce its corresponding relation

$$\textbf{def } Q \; x \; s \; =$$

$$\textbf{if } \textit{null } s$$

$$\textbf{then } \textit{nullists}$$

$$\textbf{else if } x \; = \; h \; s$$

$$\textbf{then } \textit{us}(t \; s \,)$$

$$\textbf{else } \textit{nullists}$$

If the string starts with the character x then the result is the stream $us(t \; s)$, otherwise it is *nullists,* the stream which represents the empty set. There are two special relations that correspond to: 1) the empty set whose value is always the empty set, and 2) the nullist set whose value is the set containing one member, the original argument string. These parsers are defined below.

$$\textbf{def } \textit{nullstring } s \; = \; \textit{us } s$$

$$\textbf{def } \textit{empty } s \; = \; \textit{nullists}$$

From these relations new relations can be constructed by replacing each union $(f | g)$ by the union of two relations:

$$\textbf{def } union\ f\ g\ s\ =\ appendl(f\ s)(g\ s)$$

and by replacing each Cartesian concatenation operator $(f \cdot g)$ by the concatenation of two relations.

$$\textbf{def } followedby\ f\ g\ s\ =\ concatq\,(mapl\ f(g\ s))$$
$$\textbf{where } concatq\ =\ stream1\ nullists\ appendl\ I$$

The relation $(followedby\ f\ g)$ therefore finds the set of strings related to s by f, and, for each member, finds the set related by g and forms the union of this set of sets. The context-free productions of a language without left-recursive symbols can now be reinterpreted as a set of mutually recursive functions defining the parsing relation for the language. A string belongs to the language provided that the null list is a member of the set produced by applying the parsing relation for the language to the string. In other words if L is the relation for the language, then the string is recognized if $(exists\ null\ (L\ s))$ is **true**, where $exists$ is defined below.

$$\textbf{def rec } exists\ p\ s\ =$$
$$\textbf{if } nulls\ s$$
$$\textbf{then false}$$
$$\textbf{else let } x,\ y\ =\ s()$$
$$\textbf{if } p\ x$$
$$\textbf{then true}$$
$$\textbf{else } exists\ p\ y$$

In order to produce a parser from this recognizer it is necessary to elaborate a relation so that it produces a set of pairs. The first of the pair is the object produced from the phrase recognized at the head of the string. The second of the pair is the remaining string. The $followedby$ function has now to be elaborated by providing another function as argument. The function $(cc\ h\ f\ g)$, defined below, produces the set of pairs found by applying f and g and combining the results of f and g by applying the function h.

$$\textbf{def } cc\ h\ f\ g\ s\ =$$
$$concatq\ (mapl\ q\,(f\ s))$$
$$\textbf{where } q(u,\ v)\ =\ mapl\ r(g\ v)$$
$$\textbf{where } r(y,\ z)\ =\ (h\ u\ y,\ z)$$

The $union$ operator is unchanged.

The parsing relation that produces the set of structural descriptions of a string for the language defined by the productions

$$S \rightarrow AbC$$
$$S \rightarrow Cb$$
$$C \rightarrow abS$$
$$C \rightarrow c$$
$$A \rightarrow a$$
$$A \rightarrow aC$$

can be derived in the following way. The productions are first rewritten as:

$$S = (A \cdot Q'b' \cdot C) \,|\, (C \cdot Q'b')$$
$$C = Q'a' \cdot Q'b' \cdot S \,|\, Q'c'$$
$$A = Q'a' \,|\, Q'a' \cdot C$$

and then the definitions of the basic relations are changed from the type (*string* → *string-stream*) to the type

$$string \rightarrow (A \times string) - stream$$

as follows:

$$\textbf{def } nullstring \; s \; = \; us((), s)$$
$$\textbf{def } empty \; s \; = \; nullists$$

When a character is recognized, its structural description is a tree whose root is the character and whose listing is the nullist. The definition of the parser associated with Q follows:

> **def** $Q \; x \; s =$
>> **if** *null s*
>> **then** *nullists*
>> **else if** $x = h \; s$
>>> **then** $us(ctree \; x \; (), t \; s)$
>>> **else** *nullists*

The *cc* operation can be extended to operate on a list of relations to produce a list of the objects produced from the component phrases and the remaining string.

$$\textbf{def } cclist = list1 \; nullstring \; (cc \; prefix) \; I$$

A function called '*edit*' is next introduced for changing the first member of each pair produced from a parser by applying a function f to it.

$$\textbf{def } edit\ f\ r\ s\ =\ mapl\ g\ (r\ s)\ \textbf{where } g(x, y) = (f\,x, y)$$

The function for adding a root to a listing to produce a tree is therefore

$$\textbf{def } struct\ x\ =\ (edit\ (ctree\ x))$$

and the parser for producing the structural descriptions of a string is

> **def rec**
>
> $\quad S\ =\ struct\ 'S'\ (union(cclist\ (A,\ Q'b,'\ C))(cclist\,(C,\ Q'b')))$
>
> **and**
>
> $\quad C\ =\ struct\ 'C'\ (union(cclist\ (Q'a,'\ Q'b,'\ S\,))(Q'c'))$
>
> **and**
>
> $\quad A\ =\ struct\ 'A'\ (union\ (Q'a')(cclist\ (Q'a'\ C)))$

When S is applied to a string it produces a stream of pairs. The first of each pair is the structural description of a phrase of S; the second is the remaining string.

Top-down parsing with limited backtracking. The parsers described in the previous section produce a stream of possible analyses of a string by applying (*filters* (*null* · *second*)) to the stream produced by applying L, the parsing relation, to the string. It is possible to obtain all possible analyses by translating the stream to a list. In this section the same technique of reinterpreting the productions as definitions of parsers will be used; but this time at most one analysis will be produced, and not all possible paths will be examined. Although the present method of reinterpreting descriptions of sets as descriptions of parsers employs backtracking, not all possible paths are examined; and so there is no guarantee that the parser produced in this way is a recognizer of the language. This is because the strategy that corresponds to a union such as

$$X \rightarrow a \mid ab$$

is to pick the first applicable parser and to ignore any other parser which may also be applicable. If X occurs in the production

$$Y \rightarrow Xc,$$

then the application of the parser corresponding to Y to abc will recognize a as being an instance of X, fail to find a c and so report a failure to find a Y, whereas in fact $Y \Rightarrow^* abc$, which can only be discovered by examining the second alternative for X. In spite of this difficulty the top-down partial-backup approach is popular because programs correspond to phrases and as a consequence semantic extensions can be easily made. The actions taken are formed by elaborating the program that corresponds to a phrase.

There is an important class of languages that require no backup in the string during a parse using this method; and, for these, the top-down partial backtracking method is the most convenient way to write the parser. It is possible, using this method, to intermix the parsing of context-free languages with parsing of noncontext-free phrases. In fact, since the programs that are being written are only guided by the syntax of the language, it is possible at any stage to step outside the restrictions implied by the context-free productions if this seems useful. This can be accomplished by defining parsers that depend on extra information not found in the productions and by defining new high-level functions which operate on, and produce, parsers.

A table-driven parser.   It is possible to arrange the productions in a table or tree which has the form of a two-level list for each nonterminal. The first level of the list represents a union; the second level represents Cartesian concatenation. The productions for the language are displayed in Fig. 4.7. This technique has the advantage that the program can be written once for all. It is merely necessary to change the table in order to change the language that is being processed.

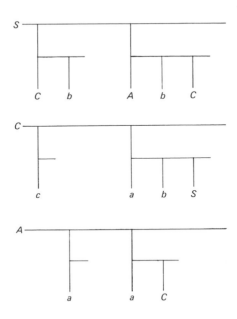

Fig. 4.7   Productions in tree form

Suppose that there is a function called *tran* which operates on a symbol, and produces its associated *(symbol-list)-list*. By convention, when *tran* is applied to a terminal symbol it produces the null list. The function *parse*, defined below, operates on a symbol and a string and produces a triple containing either **false**, (), and the original string, or **true**, the structural description of the string, and the null list.

> **def** *parse g s* =
>> **if** *null s*
>> **then false,** (), *s*
>> **else if** *null (tran g)*
>>> **then if** *g* = *h s*
>>>> **then true,** *ctree(h s)*(), *t s*
>>>> **else false,** (), *s*
>>> **else let** *b*1, *c*1, *s*1 = *union(tran g)*()*s*
>>> *b*1, *ctree g c*1, *s*1

It is defined in terms of the function *union* which takes a (symbol-list)-list as argument and tests whether the members of the list describe the string. It finds the first one that does and returns the forest of its components.

> **def rec** *union x c s* =
>> **if** *null x*
>> **then false,** (), *s*
>> **else let** *b*1, *c*1, *s*1 = *ccat (h x) c s*
>>> **if** *b*1
>>> **then** *b*1, *c*1, *s*1
>>> **else** *union (t x) c s*
>
> **and** *ccat x c s* =
>> **if** *null x*
>> **then true,** *c, s*
>> **else let** *b*1, *c*1 *s*1 = *parse (h x)s*
>>> **if** *b*1
>>> **then** *ccat(t x)(postfix c*1 *c)s*1
>>> **else false,** (), *s*

The *ccat* function accumulates the structural description in its argument *c*.

The same algorithm can be implemented by systematically deriving a program from the productions of the language, as discussed in the following section.

Parsing functions.   An alternative approach is to construct parsing programs which have the same structure as the productions of a language. There is a close correspondence between syntactic phrases and their parsers. Each phrase can be associated with a parser for recognizing whether a string starts with that phrase and produces a triple as a result. The first element of the triple is a truth value indicating whether some initial segment of the string has been recognized; the second element is something derived from the string if it is recognized; and the third is a string which is either the remainder of the initial string after the recognized segment has been removed or, if the first element is **false**, is the original string. The basic parsers can be defined as follows:

$$
\begin{aligned}
&\textbf{def } anycharacter\ s\ =\\
&\qquad \textbf{if } null\ s\\
&\qquad \textbf{then false, } (),\ s\\
&\qquad \textbf{else true, } h\ s,\ t\ s\\
&\textbf{def } empty\ s\ =\ \textbf{false, } (),\ s\\
&\textbf{def } nullistset\ s\ =\ \textbf{true, } (),\ s
\end{aligned}
$$

The parser corresponding to *anycharacter* is not applicable if the string is null. The parsers *empty* and *nullist* defined above correspond to the sets with the same name. The function called *is*, defined below, transforms a predicate on characters such as *digit, letter, vowel,* into a parser.

$$
\begin{aligned}
&\textbf{def } is\ p\ s\ =\ \textbf{let } b1, c1, s1\ =\ anycharacter\ s\\
&\qquad \textbf{if } b1\\
&\qquad \textbf{then if } p\ c1\\
&\qquad\qquad \textbf{then } b1, c1, s1\\
&\qquad\qquad \textbf{else false, } (),\ s\\
&\qquad \textbf{else false, } (),\ s\\
&\textbf{def } eqch\ x\ =\ is\ (eq\ x)
\end{aligned}
$$

The function *eqch*, which produces the parser that tests whether the first character of a string is equal to its argument, is defined above in terms of *is* and the function *eq* which tests whether two characters are equal. The operations on parsers that correspond to union and concatenated cross product are defined below as prefixed functions called *un* and *cc*.

$$\textbf{def } un \; A \; B \; s \; = \; \textbf{let } b1, c1, s1 \; = \; A \; s$$
$$\textbf{if } b1$$
$$\textbf{then } b1, c1, s1$$
$$\textbf{else } B \; s$$
$$\textbf{def } cc \; f \; A \; B \; s \; = \; \textbf{let } b1, c1, s1 \; = \; A \; s$$
$$\textbf{if } b1$$
$$\textbf{then let } b2, c2, s2 \; = \; B \; s1$$
$$\textbf{if } b2$$
$$\textbf{then } b2, f \, c1, c2, s2$$
$$\textbf{else false, } (), s$$
$$\textbf{else false, } (), s$$

The function *un* operates on two parsers, *A* and *B*, to produce the parser that first tests whether the string starts with the parser *A*. If it does the result of (*un A B*) is the result of applying *A* to the string. If not then (*un A B*) applies *B* to the string. The function (*cc f A B*) tests whether the string starts with an *A*, and is followed by a *B*. If both tests are successful the result is the triple made up of: 1) **true,** 2) the result of applying *f* to the result parts (second of the triple) of *A* and *B,* and 3) the remaining string. The action taken on recognizing a phrase is therefore embodied in the function *f*, an argument of *cc*. It is possible to produce different objects from the same language by systematically changing the parsers for the terminal symbols and for all the actions *f*. These two operations can now be extended to operate on a list of parsers as follows:

$$\textbf{def rec } unlist \; x \; = \; \textbf{if } null \; x$$
$$\textbf{then } empty$$
$$\textbf{else } un(h \; x)(unlist(t \; x))$$

or

$$\textbf{def } unlist \; = \; listl \; empty \; un \; I$$
$$\textbf{def rec } cclist \; x \; = \; \textbf{if } null \; x$$
$$\textbf{then } nullistset$$
$$\textbf{else } cc \; prefix \; (h \; x)(cclist(t \; x))$$

or

$$\textbf{def } cclist \; = \; list1 \; nullistset \; (cc \; prefix) \; I$$

The parser (*unlist x*) tests the applicability of parsers of the list x in the order of the list and applies the first one that is. The parser (*cclist x*), when applicable, produces a list of parser results as its result. In order to operate on the result part of a parser, a function called *edit* is introduced. This is defined as:

$$\textbf{def } edit \, f \, p \, s \; = \; \textbf{let } b1, c1, s1 \, = \, p \, s$$

$$\textbf{if } b1$$

$$\textbf{then } b1, f \, c1, s1$$

$$\textbf{else false, } (), s$$

This function applies its argument f to the result of parsing if a p has been recognized. A parser used to produce a structural description corresponding to the productions:

$$A \longrightarrow u_1 \, u_2 \ldots u_n$$
$$A \longrightarrow v_1 \, v_2 \ldots v_m$$

is

$$edit \, (tree \, `A")(un(cclist \, (u_1, u_2, \ldots, u_n))(cclist \, (v_1, v_2, \ldots, v_m)))$$

The parser producer corresponding to Q operates on a string x and produces the parser that tests whether the string x is an initial segment.

$$\textbf{def } Q \, x \; = \; cclist \, (map \; eqch \; x)$$

The analogous operation in which cc is replaced by *union* is called *seek x* and tests whether the string being parsed starts with one of the characters of the string x.

$$\textbf{def } seek \, x \; = \; unlist \, (map \; eqch \; x)$$

A left-recursive production will give rise to a program which will loop indefinitely. It is, however, always possible to rewrite the productions to eliminate left recursions. Another way to deal with this problem, and also to eliminate the extra levels of structural description implied by a left recursion, is to introduce a function *qualify*, defined below. The function (*qualify x y*) recognizes an x followed by as many y's as it can find in a left to right scan. It accumulates the results of parsing by association to the left, using the functional argument f.

$$\textbf{def } qualify \, f \, x \, y \, s \; =$$

$$\textbf{let } b1, c1, s1 \, = \, x \, s$$

$$\textbf{if } b1$$

$$\textbf{then } qualify1 \, f \, c1 \, s1$$

                 else false (), s

           where rec $qualify1\ f\ c\ y\ s =$

                     let $b1, c1, s1 = y\ s$

                     if $b1$

                     then $qualify1\ f(f\ c\ c1)\ y\ s1$

                     else true, c, s

If *digit* is a predicate for digits and *cnum* converts a digit into an integer, then the function *integer,* defined below, takes the longest string of digits that it can find at the head of the string and converts it to an integer.

          def $integer = qualify\ acc\ d\ d$

                     where $acc\ x\ y = 10 \times x + y$

                         and $d = edit\ cnum(is\ digit)$

The definitions of *star, perhaps, upto, exactly* and *atleast* which were introduced in section 4.4 to describe languages carry over to parsers. If | is interpreted as *un* and · is, interpreted as $(cc\ pfx)$ then the result is a parser which produces the recognized string when successfully applied.

The parsers that correspond to the sets produced by mapping from another set are difficult to derive in any systematic way. There is, however, a useful case in which the set of strings has an initial segment, and is followed by a phrase that depends in some way on the segment found. The function which is useful in such situations is called *unmap,* and is defined below.

          def $unmap\ g\ p\ s =$

                     let $b1, c1, s1 = p\ s$

                     if bl

                     then $g\ c1\ s1$

                     else false, (), s

The function *unmap* operates on a parser p and a parser-producing function g. The parser p is first applied and then g operates on the result of p to produce another parser which is then applied to the remainder of the string. As an example, the parser $(unmap\ Q\ (Q\ x))$ tests whether the string starts with the string x duplicated. The function *unmap* corresponds to a parsing strategy in which information is gathered at the beginning of a string in order to predict the syntax of the remaining string. A second example shows its value in parsing bracketed phrases. Suppose that there is a set of opening brackets called *opens,* and a function called *mate* which maps each opening bracket to its corresponding closing bracket. The bracketing parser in which phrases are enclosed in mated brackets can be defined as follows:

$$\textbf{def } bracket\ x\ =\ unmap\ f\ (seek\,(opens))$$
$$\textbf{where } f\,y\ =\ cc\ 1st\ x\ (eqch(mate\ y))$$

As a third example of the use of *unmap,* consider a parser for the set of phrases *p,* separated by . or , or ; or : in which the same separator is used throughout so that *a,b,c, a;b;c, a.b.c,* and *a:b:c* are permitted, but mixtures such as *a;b,c* are not. A parser for these phrases can be defined in terms of *unmap* as follows:

$$\textbf{def } separated\ h\ s\ p\ =\ qualify\ h\ p\,(unmap\,(\lambda y.\ separated\ h(u\ y)p\,)seek\ s\,)$$

The parser (*separated h s p*) parses strings of phrases *p* separated by characters that belong to the set *s.* After the first separator *y* is found, the remainder of the string is parsed with the set *s* restricted to the single separator *y.* Productions for which the corresponding parser requires no backup, are considered next.

No backup parsing.   The parsers can be classified according to the maximum number of characters at the head of the string which must be examined to decide whether the parser is inapplicable. The parsers *empty* and *nullistset* look at no characters, and the parsers corresponding to terminal symbols look at one character. The backup in the string occurs when a false exit is taken from a parser. A no-backup parser is defined to be one that successfully recognizes all strings of the language, and rejects nonstrings. The only parsers that fail when recognizing a string of the language are 1-character backup parsers. Clearly, for such a parser any exit from an n-character backup parser with n > 1 should cause immediate rejection of the whole string. If the set of alternatives includes *nullistset* it should be placed last since it never fails, and any parser placed after it will never be applied. More generally there are parsers that never take a false exit which are called *nonfalse* parsers. Examples are

$$nullistset\ s\ =\ \textbf{true, }(),\ s$$

which always returns **true**, and

$$qualify1\ f\ c\ y$$

which recognizes any number of occurrences of the phrase *y* at the head of a string, including zero occurrences, and any parser made up of a union which includes a nonfalse parser. No parser which follows a nonfalse parser in a union can ever be applied. It has been shown that, if the productions of a context-free language are reinterpreted as parsers in this way, then the question of whether the resulting program will recognize precisely all strings of the language is unsolvable.

It is possible to determine whether the parser corresponding to a grammar will require no backup by examining the productions. Suppose all productions are put into the standard form

$$X \to Y_1 \mid Y_2 \mid \ldots \mid Y_m \mid Z_1 Z_2 \ldots Z_n$$

in which $m, n > 0$ and the Y's and Z's are either terminal or nonterminal symbols. If $n = 0$, $Z_1 Z_2 \ldots Z_n$ is to be regarded as the nullistset. According to Knuth [4–31] there are four conditions which are necessary and sufficient for the validity of the no-backup method, provided there are no useless nonterminal symbols.

1. The grammar has no left-recursive nonterminal symbols.
2. The sets $first(Y_1)$, $first(Y_2) \ldots first(Y_m)$, $first(Z_1 Z_2 \ldots Z_m)$ are mutually disjoint.
3. If $Z_1 Z_2 \ldots Z_n \Rightarrow e$, then $first(Y_j)$ contains no symbols in common with the set of symbols which can follow the phrase X in the language.
4. $Y_1, Y_2, \ldots, Y_m$, are not nonfalse.

A nonterminal symbol X is defined to be nonfalse if and only if either: 1) Y_j is nonfalse for some j $(1 < j < m)$, or 2) $n = 0$ or 3) $n > 0$ and Z_1 is nonfalse.

Top-down parsing using labels. The parsers just described are functions which return as part of their result a success/failure indication which then has to be tested by the calling function. It is perhaps more efficient to write programs which have a two-way branch, one for success and the other for failure. In this section the same parsers are redefined so that each is given an additional argument which specifies the action to be taken when it fails to find its corresponding phrase at the head of the string. This technique permits more flexibility in the structure of the program by breaking away from the hierarchical subroutine call and return format. Each parsing function is given an additional argument which is a basic function, closure, or program closure and which will be applied to the string when the phrase corresponding to the parser is not recognized. The result of a parser is a pair whose first is the object produced from the string and whose second is the remaining string. The simplest parsers are defined as follows:

> **def** *empty E s* = *E s*
>
> **def** *nullistset E s* = (), *s*
>
> **def** *anycharacter E s* =
>
> > **if** *null s*
> >
> > **then** *E s*
> >
> > **else** *h s, t s*

> **def** *is E p s =*
>> **let** $b1, s1 =$ *anycharacter E s*
>> **if** *p b*1
>> **then** *b*1, *s*1
>> **else** *E s*

The function from parsers to parsers corresponding to union takes the following simple form

$$\textbf{def } \textit{un f g E s} = f(g\ E)\ s$$

In other words the union function is the functional composition of *f* and *g*. First *f* is applied to the string, and if it fails (*g E*) is applied.

The function from parsers to parsers corresponding to the Cartesian concatenation operator is more complex and uses the jump operator **J** described in Chapter 2. It is used to make sure that, when either of the argument parsers *f* and *g* fail, an exit is made from the parser (*cc h f g E*) even when *E* is not a program closure.

> **def** *cc h f g E s =*
>> **let** $E1 = \textbf{J}\ E$
>> **let** $c1, s1 = f\ E1\ s$
>> **let** $c2, s2 = g\ E2\ s1$
>>> **where** $E2(s1) = E1\ s$
>>> *h c*1 *c*2, *s*2

This technique permits the action of a parser to be terminated abruptly upon failure so that the action continues at some level higher than the point at which the parser was called.

Other useful methods of combining parsers are

> **def** *difference f p E s =* **let** $b1, s1 = f\ E\ s$
>> **if** *p b*1
>> **then** *E s*
>> **else** *b*1, *s*1

which fails if either *f* is inapplicable or if the result of parsing satisfies the predicate *p*. Thus

stringnotcontainingsemicolon = star (*difference anycharacter* (*eq*';'))

Redefinitions of the other parsers follow:

$$\textbf{def } qualify \ h \ f \ g \ E \ s$$
$$= \ qualify1 \ h \ g \ (f \ E \ s)$$
$$\textbf{where } qualifyl \ h \ g \ (b, s)$$
$$= \textbf{let } E1 = \textbf{J} \ (\lambda s1.b, s)$$
$$\textbf{let } b1, s1 = g \ E1 \ s$$
$$qualify1 \ h \ g \ (h \ b \ b1, s1)$$

$$\textbf{def } unmap \ f \ p \ E \ s =$$
$$\textbf{let } E = \textbf{J} \ E$$
$$\textbf{let } b1, s1 = p \ E \ s$$
$$f \ b1 \ E \ s1$$

$$\textbf{def } edit \ g \ f \ E \ s =$$
$$\textbf{let } b1, s1 = f \ E \ s$$
$$g \ b1, s1$$

4.6 LEFT-CORNER BOTTOM-UP PARSING

The left-corner bottom-up parsing method is so called because the structural description tree is constructed from the left corner first. The parsing program continually finds itself in a situation where it has found a certain phrase X as the initial segment of a string and is attempting to find a *goal* phrase G. It therefore seeks a continuation which when it follows an X, produces a G. This strategy also has a nondeterministic pushdown automaton associated with it.

A pushdown automaton for left-corner recognition.   There are three types of transitions of the automaton, the first of which is

$$(G, U_1) \rightarrow (G \ X / U_n \ U_{n-1} \dots U_2, ()) \tag{i}$$

for every nonterminal symbol G and production

$$X \rightarrow U_1 \ U_2 \ U_3 \dots U_n.$$

The symbol / is a special symbol which is neither terminal nor nonterminal. The second type of transition is

$$(B, B) \rightarrow ((), ()) \tag{ii}$$

for every terminal or nonterminal B of the language. The third is

$$(A /, ()) \rightarrow ((), A) \tag{iii}$$

for every nonterminal A.

The strategy of this pushdown automaton can be described informally as follows. When a state $(\ldots G, U_1 \ldots)$ has been reached it indicates that the automaton is looking for a G at the head of the string, and has found a U_1. The transition to the state

$$(G\ X/U_n\ U_{n-1}\ldots U_2,\ ())$$

switches its attention to look for a U_2, followed by a $U_3 \ldots$ and a U_n in succession. The symbol X is stored on the pushdown list to indicate that if it discovers that the head of the string matches $U_2\ U_3 \ldots U_n$ then it has found an X. The transformation

$$(G\ X/,\ ()) \rightarrow (G,\ X)$$

made at this stage records the fact that is has found an X when looking for a G. It follows that the number of state transitions of the type shown in eq.(i) can be reduced by requiring that only transitions with the property that $X\ \varepsilon\ left^*\ G$ are included. This is because if X does not belong to $left^*\ G$ then the state $(G,\ X \ldots)$ cannot lead to the state $((),\ ())$. The pushdown automaton is started in the state $(S,\ x)$ where x is the string to be recognized, and S is the symbol for the language. If a sequence of transitions lead to the state $((),\ ())$ then the string has been recognized. The state transitions corresponding to the productions

$$S \rightarrow AbC$$
$$S \rightarrow Cb$$
$$C \rightarrow abS$$
$$C \rightarrow c$$
$$A \rightarrow a$$
$$A \rightarrow aC$$

are given in Fig. 4.8. The steps taken when recognizing the string '*aabcbbc*' are shown in Fig. 4.9.

(a) $(S,A) \rightarrow (SS/Cb, ())$
$(S,C) \rightarrow (SS/b, ())$
$(S,a) \rightarrow (SA/, ())$
$(S,a) \rightarrow (SA/C, ())$
$(S,a) \rightarrow (SC/Sb, ())$
$(S,c) \rightarrow (SC/, ())$
$(A,a) \rightarrow (AA/, ())$
$(A,a) \rightarrow (AA/C, ())$
$(C,a) \rightarrow (CC/Sb, ())$
$(C,c) \rightarrow (CC/, ())$
(b) $(X,X) \rightarrow ((), ())$ for $X = S,A,C,a,b,c$
(c) $(Y/,()) \rightarrow ((),Y)$ for $Y = S,A,C$

Fig. 4.8   Example of transitions of a left-corner pda

Pushdown	String	Transition
S	aabcbbc	(S,a) → (SA/C,())
SA/C	abcbbc	(C,a) → (CC/Sb, ())
SA/CC/Sb	bcbbc	(b,b) → ((),())
SA/CC/S	cbbc	(S,c) → (SC/, ())
SA/CC/SS/b	bbc	(b,b) → ((),())
SA/CC/SS/	bc	(S/,()) → ((),S)
SA/CC/S	Sbc	(S,S) → ((),())
SA/CC/	bc	(C/,()) → ((),C)
SA/C	Cbc	(C,C) → ((),())
SA/	bc	(A/,()) → ((),A)
S	Abc	(S,A) → (SS/Cb,())
SS/Cb	bc	(b,b) → ((),())
SS/C	c	(C,c) → (CC/,())
SS/CC/	()	(C/,()) → ((),C)
SS/C	C	(C,C) → ((),())
SS/	()	(S/,()) → ((),S)
S	S	(S,S) → ((),())
()	()	

Fig. 4.9   Steps in the recognition by a left corner bottom-up pda

Limited backtrack left-corner parsing. In left-corner bottom-up parsing it is convenient to factorize the right hand side of productions, and store them as a symbol-forest. In this forest the productions with a common head symbol are arranged as a tree whose root is that symbol, and whose listing is made up of trees formed in the same way from the tails of the productions having a common head. At each end point of this tree is stored the symbol defined by the production found on the path leading from the root to the end point. The productions

$$S \rightarrow Cb \mid AbC$$
$$C \rightarrow c \mid abS$$
$$A \rightarrow a \mid aC$$

for example, are put in the form illustrated in Fig. 4.10.

Suppose there is a function called *tran* from each symbol to the forest of its continuations. After a phrase has been found, the number of possible continuations to be examined can be reduced by using a table for a relation

Fig. 4.10   Forest of produtions for left-corner parsing

succr. If (*succr p q*) is **true** then $p \varepsilon \ left^* q$. The table for the productions above is

	S	C	A
S			
C	√		
A	√		
a	√	√	√
b			
c	√		

The following program, called *parse,* tests whether a string belongs to the language *g* and, if it does, produces the structural description of the string in the form of a symbol-tree.

> **def** *parse g s* =
> > **if** *null s*
> > **then false,** (), *s*
> > **else** *recognize*(*h s*)()*g*(*t s*)

The main work of parsing is embodied in a function called *recognize.* The form of the function *parse* implies that the productions cannot contain the nullistset. The function (*recognize x c g*) operates on the symbol *x* which has already been recognized, the symbol-forest *c* which is the structural description of the string recognized so far, and *g*, the eventual goal. It produces either **false**, nullist, and the original string, or **true**, the structural description tree with root *g*, and the remaining string.

> **def** *recognize x c g s* =
> > **if** *succr x g*
> > **then let** *b*1, *c*1, *s*1 = *diagram* (*tran x*)(*u*(*ctree x c*))*g s*
> > > **if** *b*1
> > > **then** *b*1, *c*1, *s*1
> > > **else if** *x* = *g*
> > > > **then true,** *ctree g c, s*
> > > > **else false,** (), *s*
> > **else if** *x* = *g*
> > > **then true,** *ctree g c, s*
> > > **else false,** (), *s*

The main burden in *recognize* is put on the function *diagram*. The function *recognize* is arranged so that a possible continuation can occur even when $x = g$, and so it finds the longest phrase g at the head of the string when there is a left-recursive production. The diagram function operates on x, a symbol forest which describes the possible continuations; c, a symbol-forest representing the structural description found so far; g, the goal symbol; and s, the string to be recognized.

> **def rec** *diagram x c g s* =
>> **if** *null x*
>> **then false,** (), *s*
>> **else let** $b1, c1, s1 = diag(h\ x)\ c\ g\ s$
>>> **if** $b1$
>>> **then** $b1, c1, s1$
>>> **else** *diagram* $(t\ x)\ c\ g\ s$

The *diagram* function tests for the union of the possible continuations and is defined in terms of *diag*, which operates on a symbol-tree and tests for the concatenations of the continuations.

> **def** *diag x c g s* =
>> **if** *null*(*listing x*)
>> **then** *recognize* (*root x*) *c g s*
>> **else**
>>> **let** $b1, c1, s1 = parse\ (root\ x)\ s$
>>> **if** $b1$
>>> **then** *diagram* (*listing x*)(*postfix* $c1\ c$)*g* $s1$
>>> **else false,** (), *s*

If the symbol is a terminal node then it means that the phrase described by that symbol has been found, and the next step is to complete the recognition of the goal g. Otherwise an attempt is made to recognize the root node followed by one of the symbols in the listing.

The parser (*recognize x c g*) is one which is used after a phrase x has been found. It attempts to add to x to form a phrase of g.

The productions can be transformed so that when the top-down strategy is applied to the transformed productions essentially the same steps are followed as if the left-corner bottom-up strategy had been used on the original productions. To do this, new phrases of the type [x:g] are introduced which represent the set of strings which make a phrase g when concatenated to the end of a phrase of x. The top-down parser $[x:g]$ is the function (*recognize x c g*). The productions can be transformed as follows. First write them in a matrix form

$$(S,\ C,\ A) = (S,\ C,\ A) \begin{pmatrix} \phi\ \phi\ \phi \\ b\ \phi\ \phi \\ bC\ \phi\ \phi \end{pmatrix} + (\phi,\ abS + c,\ a + aC)$$

i.e. $\mathbf{x} = \mathbf{x}\,G + f$ in which $\mathbf{x}$ is a vector of nonterminal symbols. Multiplication stands for Cartesian concatenation and addition stands for union. The rightmost symbols of f are terminal symbols. The equations can be solved for $\mathbf{x}$ giving

$$\mathbf{x} = fG^* \text{ where } G^* = I + GG^*$$

where I is a matrix with diagonal elements equal to e (the nullistset) and off-diagonals equal to ϕ (the empty set). This solution is the new set of productions. The elements of G^* can be indexed by pairs of original nonterminal symbols to produce new nonterminal symbols of the form $[A:B]$. The new productions corresponding to $\mathbf{x} = fG^*$ are formed from the old productions whose right-hand side starts with a terminal symbol. Thus

$$(S,\ C,\ A) = (\phi,\ abS + c,\ a + aC)\, G^*$$

gives

$$S = (abS + c)\,[C:S] + (a + aC)\,[A:S]$$
$$C = (abS + c)\,[C:C] + (a + aC)\,[A:C]$$
$$A = (abS + c)\,[C:A] + (a + aC)\,[A:A]$$

These can be reduced because a symbol $[X:Y]$ is empty if X does not belong to *left* * Y. It follows that $[A:C]$ and $[C:A]$ are empty, leaving the productions:

$$S \rightarrow abS\,[C:S]$$
$$S \rightarrow c\,[C:S]$$
$$S \rightarrow a\,[A:S]$$
$$S \rightarrow aC\,[A:S]$$
$$C \rightarrow abS\,[C:C]$$
$$C \rightarrow c\,[C:C]$$
$$A \rightarrow a\,[A:A]$$
$$A \rightarrow ac\,[A:A]$$

The other equation $G^* = I + GG^*$ defines the new nonterminal symbols

$$G^* = \begin{pmatrix} e\,\phi\,\phi \\ \phi\,e\,\phi \\ \phi\,\phi\,e \end{pmatrix} + \begin{pmatrix} \phi\,\phi\,\phi \\ b\,\phi\,\phi \\ bC\,\phi\,\phi \end{pmatrix} G^*$$

The first term gives rise to the productions

$$[S:S] \rightarrow e$$
$$[C:C] \rightarrow e$$
$$[A:A] \rightarrow e.$$

The second term produces a new production of the form

$$[C:A] \rightarrow U_1\,U_2 \ldots U_n\,[B:A]$$

for every original production of the form

$$B \rightarrow C\,U_1\,U_2 \ldots U_n$$

in which C, the first symbol, is nonterminal, and A is a nonterminal such that $C \; \varepsilon \; left^*A$.

Thus from the productions

$$S \rightarrow Cb$$
$$S \rightarrow AbC$$

are derived the productions

$$[C:S] \rightarrow b[S:S]$$
$$[a:S] \rightarrow bC[S:S]$$

The original nonterminals are defined by productions having a terminal symbol as the first symbol on the right-hand side. The productions for the new symbols have an original symbol as first symbol on the right-hand side. It follows that the productions can be put in a form in which every product with a nonempty right side has a first symbol which is a terminal symbol. This is done by substituting into all productions for new symbols whose first symbol is nonterminal.

The new set of productions are

$$S \rightarrow abS[C:S]$$
$$S \rightarrow c[C:S]$$
$$S \rightarrow a[A:S]$$

$$S \rightarrow aC[A:S]$$
$$C \rightarrow abS$$
$$C \rightarrow c$$
$$A \rightarrow a$$
$$A \rightarrow aC$$
$$[C:S] \rightarrow b$$
$$[A:S] \rightarrow bC.$$

The top-down strategy used on these productions produces the structural description in Fig. 4.11 when applied to the string *aabcbbc*.

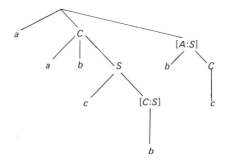

Fig. 4.11 Top-down parsing tree using transformed productions

4.7 RIGHT-CORNER BOTTOM-UP PARSING

In this case the derivation tree is constructed from the right corner first in a left-to-right scan of the string. A pushdown list is used to record the analysis of the string that has been examined. The steps are: 1) to move a character from the input to the top of the pushdown list, and 2) to try to find a right side of a production at the head of the pushdown and replace it by its left side. The strategy is therefore to continually look backward at the already-analyzed initial segment and try to build the derivation tree.

Pushdown automaton. There is a nondeterministic pushdown automaton for the right-corner bottom-up strategy of parsing. It has two types of transition. For every production

$$A \rightarrow U_1 U_2 \ldots U_n$$

there is a state transition

$$(U_1 U_2 \ldots U_n, ()) \rightarrow (A, ()),$$

in which the right side of the production is found at the head of the push-down and is replaced by its left-hand side. This is called a *reduction step* or a *reduce operation*. The second type of transition takes the form

$$((), A) \rightarrow (a, ())$$

for every terminal symbol *a*, and is called a *shift*. In this step a symbol is moved from the head of the string to the head of the pushdown. If the bottom-up pda is started in the state

$$(), x$$

where *x* us the string to be recognized, and if any sequence of transitions can be found that leads to the state

$$S, (),$$

then the string has been recognized as belonging to the language *S*. The transitions for the same productions are given in Fig. 4.12. The sequence of states that recognizes the string *aabcbbc* is given in Fig. 4.13.

$S \rightarrow AbC$	$(AbC, ()) \rightarrow (S, ())$	(1)
$S \rightarrow Cb$	$(Cb, ()) \rightarrow (S, ())$	(2)
$C \rightarrow abS$	$(abS, ()) \rightarrow (C, ())$	(3)
$C \rightarrow c$	$(c, ()) \rightarrow (C, ())$	(4)
$A \rightarrow a$	$(a()) \rightarrow (A, ())$	(5)
$A \rightarrow aC$	$(aC, ()) \rightarrow (A, ())$	(6)
	$((), a) \rightarrow (a, ())$	(7)
	$((), b) \rightarrow (b, ())$	(8)
	$((), c) \rightarrow (c, ())$	(9)

Fig. 4.12   The transitions for a right-corner pda

Pushdown list, head at right	Input string, head at left	Transformation number
()	aabcbbc	7
a	abcbbc	5
aa	bcbbc	8
aab	cbbc	9
aabc	bbc	4
aabC	bbc	8
aabCb	bc	2
aabS	bc	3
aC	bc	6
A	bc	8
Ab	c	9
Abc		4
AbC		1
S		

Fig. 4.13   Steps in right-corner pda recognition

Since the production that is reduced is the one at the top of the push-down, the order in which productions are reduced is the reverse of a right-most derivation of the string. One way to organize the work to be done by the pushdown automaton is to represent the set of pushdown lists which are the possible structural descriptions of an initial segment of the string as a symbol-forest in which both the heads and tails are shared.

The program alternately adds a new character to the forest and then reduces it by replacing any path from the root which is the right side of a production by its left-side symbol. It is convenient to store the productions in a forest in reverse order so that the tail segments of the right-hand side are shared. The rules

$$S \rightarrow AbC \mid Cb$$
$$C \rightarrow abS \mid c$$
$$A \rightarrow a \mid aC$$

would then be stored as in Fig. 4.14, in which the end points are the left hand sides of productions. The analysis of the string *aabcbbc* proceeds as in Fig. 4.15.

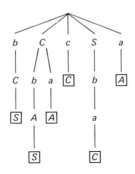

Fig. 4.14 Forest of productions for right-corner parsing

The root to leaf paths through the resulting forest are all possible analy-ses of the string. Since S is at the top level and is an end point, the string belongs to language S.

The following function produces the set of trees that immediately depend on a forest y according to the productions x arranged as a symbol-

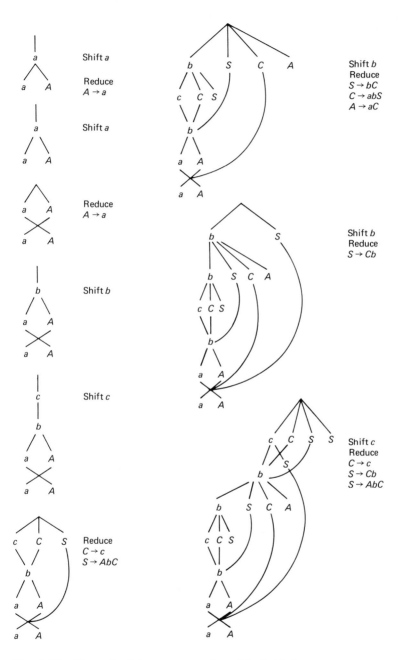

Fig. 4.15   Stages in right-corner parsing of *'aabcbbc'*

forest such as that in Fig. 4.14.

def rec *trees x y =*
    if *null x*
    then ()
    else *concat* (*try*(*h x*)*y*) (*trees* (*t x*) *y*)
  where rec *try x y =*
    if *null y*
    then ()
    else if *null* (*listing x*)
      then *u* (*ctree* (*root x*) *y*)
      else if *root x* = *root* (*h y*)
        then *trees* (*listing x*)(*listing* (*h y*))
        else *try x*(*t y*)

This function is then iterated until no further reductions by the function *reduce* are possible. Reduce is defined as:

def rec *reduce x y =*
    let *p* = *trees x y*
    if *null p*
    then ()
    else *union p* (*reduce x p*)

The forest *y* is then transformed by alternately shifting and reducing by the following function

parse x y s =
    if *null s*
    then *y*
    else let *z* = *u*(*ctree* (*h s*)*y*)
      parse x (*union x* (*reduce x z*))(*t s*)

Infixed operator expressions.   Although the applicative structure of an infixed operator expression like

$$a + b = (c + d) - e \,/\, f(x)$$

can be indicated by productions, it is simpler to introduce a notation of the "precedence strength" of the unfixed operators. The most frequently used rules can be summed up in the acronym BODMAS, which stands for

Brackets, Of, Division, Multiplication, Addition, and Subtraction. This indicates that bracketed expressions should be evaluated as a unit first, followed by application of function to argument ($f(x)$ can be read as the f of x), division, multiplication, addition, and subtraction in that order. Usually addition and subtraction are given the same strength and, to specify the order of evaluation precisely, rules must be given for dealing with expressions such as

$$a + b - c + d.$$

whose operators have equal precedence.

The rules for ALGOL 60 can be given by attaching a strength to each operator as follows:

3	$\equiv$
4	$\supset$
5	$\vee$
6	$\wedge$
7	$\neg$
8	$\leq, <, \neq, \geq, >, =$
9	$+, -$
10	$\times, /, \div, neg$
11	$\uparrow$
12	*application*

The rule in ALGOL 60 is to break ties by grouping to the left so that $a + b - c + d$ is bracketed $((a + b) - c) + d$.

An infixed operator expression can be translated to reverse Polish notation by using a pushdown list. The identifiers are passed through from the input to the output string. When an operator is at the head of the string its strength is compared with the strength of the operator at the head of the pushdown list. If the operator at the head of the string is stronger then it is added to the pushdown list; if it is weaker the operator at the head of the pushdown is transferred to the output and the head of the input string is tested against the new head of the pushdown in the same way.

The steps in the translation of $(a + b \times c + d)$ to reverse Polish notation are shown in Fig. 4.16.

The initial pushdown list must contain an operator which is weaker than any other in the string. This is represented by (in Figure 4.16. The string must also end with a weak operator ()) to clear the operators in the pushdown down to (. When two matching parentheses are compared, they are both removed.

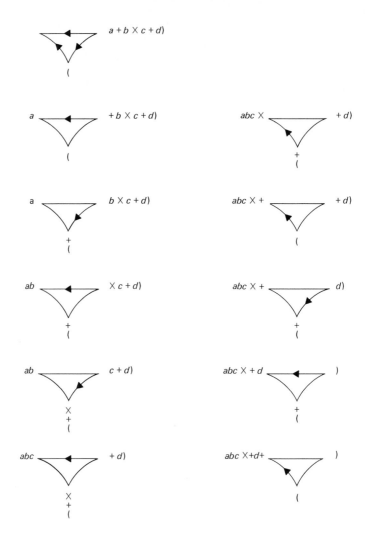

Fig. 4.16 Translating expressions to reverse Polish

It seems to be a general rule that whenever a pushdown list appears explicitly in a program it is possible to rewrite the program in a functional notation without an explicit pushdown list. This is true in this case. Suppose a parsing program for infixed operator expressions has two arguments: 1) *ops,* the set of infixed operators allowed in strings, and 2) *p,* the type of phrase that is separated by infixed operators. The infixed operator expression can be grouped as follows. It is made up of the phrase *p* followed

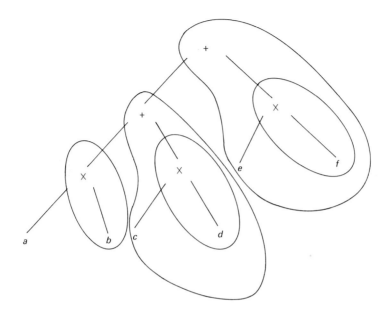

Fig. 4.17   Grouping of an infixed-operator string

by a list of groups. A group in turn is made up of an infixed operator *o* followed by an infixed operator expression containing operators whose precedence is stronger than *o*. The infixed operator string $a \times b + c \times d + e \times f$ would be grouped as in Fig. 4.17 in which each list of groups forms the leftmost edge of a tree.

The function that performs an analysis in this way is called *infop*. Its arguments are: *p*, the parser for the phrase to be separated; *ops*, the set of permitted infixed operators; and *f*, a function for accumulating results of parsing. The function also depends on a relation called *stronger* between infixed operators. The definition of *infop* is:

> **def rec** *infop f ops* =
>
>   *qualify f p* (*unmap* ((*infop f p*) • *stronger*)(*seek ops*))

The parser first looks for a phrase using the parser *p* and, if one is found, next looks for a list of groups; if *p* is not applicable, (*infop f p ops*) is not applicable either. The parser (*seek ops*) tests whether a group begins with an infixed operator *o* belonging to *ops*. If it finds one that does, then the parser (*infop f p* (*stronger o*)) is applied to the remainder of the string. This is a parser for infixed operator expressions containing operators stronger than *o*. The function *f* accumulates the results of parsing groups. It seems best to add another functional argument g which is applied to the infixed operator and the remainder of a group. The parser can then be redefined as follows

> **def rec** *infop f g p ops =*
>
> *qualify f p (unmap group (seek ops))*
>
> **where** *group x =*
>
> *edit (g x)(infop f g p (stronger x))*

Suppose that p is a parser for an identifier whose value is itself, i.e., *identifier = qualify prefix letter(un letter digit)* then:

1. the string can be bracketed by using $f\ x\ y\ =\ =\ concat$ ('(', x, y, ')') and $g = prefix$;
2. a binary tree can be constructed by using $f(x, y) = x, y$ and $g\ x\ (y, z) = cbtree\ y(cbtree\ x\ etree\ etree)\ z$;
3. a reverse Polish string can be constructed by using $g = postfix$ and $f = concat$;
4. the expression can be evaluated by using $g\ x\ y = x, y$ and $f\ x\ (y, z) = apply\ (value\ E\ y)(value\ E\ x, z)$;
5. infixed operator strings with brackets can be parsed by specifying p to be a bracketed expression; e.g.,

> *infop f g(bracket (infop f g p ops))*
>
> **where**
>
> *bracket x = cc 2nd (Q'(')(cc 1st x(Q')')*
>
> **where** $1st\ x\ y = x$
>
> **and** $2nd\ x\ y = y.$

Simple precedence grammars. In order to produce an efficient parsing strategy from the general right-corner bottom-up strategy there must be a method for reducing the number of pushdown lists which appear in the analysis, preferably to one. It is possible to do this by deciding whether to perform a reduce or shift operation by examining the top symbol on the pushdown list and the top symbol on the input string. Suppose that at most one of the three relations, $<$, $\doteq$, or $>$ exists between any two symbols, that $<$ and $\doteq$ both indicate a shift operation, and that $>$ indicates a reduce operation. If a reduce operation is decided upon; one must determine how many symbols are to be reduced, and what is to replace them. The number of symbols that are candidates for a reduction are those up to the first $<$

found on the pushdown. These three relations can be obtained from the productions:

1. $A \lessdot B$ if there is a production of the form $D \rightarrow \ldots AC \ldots$ and $C \Rightarrow^+ B \ldots$
2. $A \doteq B$ if there is a production $D \rightarrow \ldots AB \ldots$
3. $A \gtrdot B$ if there is a production $D \rightarrow \ldots EF \ldots$ and $E \Rightarrow^+ \ldots A,\ F \Rightarrow^+ B \ldots$.

The parser determines when a reduce operation should be made by comparing the symbol at the head of the pushdown (A) and the symbol at the head of the string (B). If they are unrelated by any of the relations $\lessdot$, $\doteq$, or $\gtrdot$; then the string does not belong to the language. If a sequence of symbols are related as follows:

$$A \lessdot B \doteq C \doteq D \gtrdot E$$

then, upon finding that $D \gtrdot E$, a reduce operation is called into play which attempts to find a production whose right-hand side is BCD. If one is found, e.g., $F \rightarrow BCD$, then the string BCD is replaced by F and comparisons are started again by comparing A and F.

The precedence-grammar parser can be expressed as a function which has an error action argument E which is invoked whenever either no relation exists between two symbols or the reduce function is unable to find a production whose right side is equal to the current candidate.

The second argument y is the structural description of a segment (e.g. the $B \doteq C \doteq D$ in $\ldots A \lessdot B \doteq C \doteq D \ldots$)

```
def rec prec E y s =
    if null s
    then reduce E y
    else if h y < h s
        then let b1, s1 = prec E (u(hs))(t s)
                 prec E y s1 (b1:s1)
        else if h y = h s
            then prec E (h s:y)(t s)
            else if h y > h s
                then prec E (u(reduce E y))s
                else E s
```

The productions

$$A \rightarrow B$$
$$B \rightarrow BaC$$
$$B \rightarrow C$$
$$C \rightarrow bC$$
$$C \rightarrow b$$

give rise to the relations expressed in table form in Fig. 4.18. The steps in the parsing of the string *babb* according to these productions are shown in Fig. 4.19.

	A	B	C	a	b
A					
B				$\doteq$	
C				$\cdot >$	
a		$\doteq$			$< \cdot$
b		$\doteq$	$\cdot >$		$< \cdot$

Fig. 4.18 Precedence relations

The function *prec E(u(h s))(t s)* is applied to the strings. In other words, the analysis starts with a shift operation. The string is therefore recognized as belonging to the language *B*. The resulting structural description is:

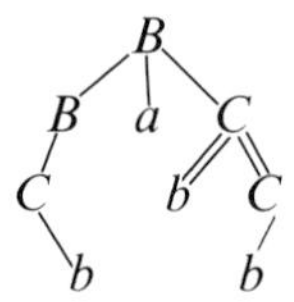

A grammar is called a *simple precedence grammar* if, and only if, no more than one of the relations $>$, $\doteq$, or $<$ holds between any two symbols, and if the grammar has no common right parts in its productions.

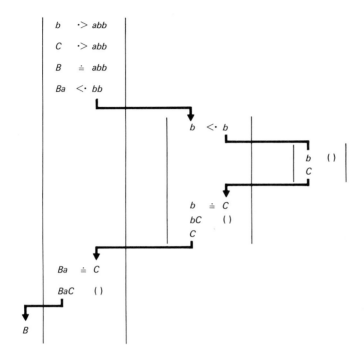

Fig. 4.19   Steps in analyzing a simple precedence language

REFERENCES

4-1. Aho, A. V., and J. D. Ullman, *The Theory of Parsing, Translation and Compiling,* Englewood Cliffs, N.J.: Prentice-Hall, 1973.

4-2. Bar Hillel, Y., *Language and Information,* Reading, Mass.: Addison-Wesley, 1964.

4-3. Bobrow, D. G., "Syntactic analysis of English by computer—a survey," *Proc. AFIPS FJCC,* Vol. 24 (1963), pp. 365–387.

4-4. Brooker, R. A., and D. Morris, "The compiler-compiler," *Ann. Rev. in Automatic Programming,* Vol. 3, Oxford: Pergamon Press, 1963, pp. 229–275.

4-5. Brzowski, J. A., "A survey of regular expressions and their applications," *IRE Trans. EC-11,* June, 1962, pp. 324–335.

4-6. Brzowski, J. A., and E. J. McCluskey, "Signal flow graph techniques for sequential circuit state diagrams," *IEEE Trans. EC-12,* No. 2, Apr. 1963, pp. 67–76.

4-7. Cheatham, T. E., *The Theory of Construction of Compilers,* Computer Associates, Inc., 1967.

4-8. Cheatham, T. E., and K. Sattley, "Syntax directed compiling," *Proc. AFIPS SJCC,* Vol. 25, Spartan Books, 1964, pp. 31–57.

4-9. Chomsky, N., "Three models for the description of language," *IRE Trans. I.T.-2,* No. 3, 1956, pp. 113–124.

4-10. Chomsky, N., *Syntactic Structures,* The Hague, Holland: Mouton Co., 1959.

4-11. Chomsky, N., and M. P. Schultzenberger, "The algebraic theory of context-free languages," (in) *Computer Programming and Formal Systems,* P. Braffort and D. Hirschberg (eds.), Amsterdam: North-Holland, 1963, pp. 118–161.

4-12. Cohen, D. J., and C. C. Gotlieb, "A list structure form of grammars for syntactic analysis," *Comp. Surveys,* Vol. 2, No. 1, 1970, pp. 65–82.

4-13. Colmerauer, A., "Total precedence relations," *JACM,* Vol. 17, No. 1, 1970, pp. 14–30.

4-14. Earley, J., "An efficient context-free parsing algorithm," Ph.D. Thesis, Carnegie-Mellon U., 1968. *CACM,* Vol. 13, No. 2, 1970, pp. 94–102.

4-15. Feldman, J. A., "A formal semantics for computer languages and its application in a compiler-compiler," *CACM,* Vol. 9, No. 1, 1966, pp. 3–9.

4-16. Feldman, J. A., and D. Gries, "Translator writing systems," *CACM,* Vol. 11, No. 2, 1968, pp. 77–113.

4-17. Floyd, R. W., "A descriptive language for symbol manipulation," *JACM,* Vol. 8, No. 4, 1961, pp. 579–584.

4-18. Floyd, R. W., "The syntax of programming languages—a survey," *IEEE Trans. EC-13,* Vol. 4, 1964, pp. 346–353.

4-19. Floyd, R. W., "Syntactic analysis and operator precedence," *JACM,* Vol. 10, No. 3, 1963, pp. 316–333.

4-20. Ginsburg, S., *The Mathematical Theory of Context-Free Languages,* New York: McGraw-Hill, 1966.

4-21. Glennie, A., *On the Syntax Machine and the Construction of a Universal Compiler,* Tech. Rep. No. 2., Computation Center, Carnegie-Mellon U., 1960.

4-22. Gray, J. N., and M. A. Harrison, "Single pass precedence analysis," *IEEE Conference Record of Tenth Annual Symposium on Switching and Automata Theory,* 1969, pp. 106–117.

4-23. Greibach, S., "A new formal form theorem for context-free grammars," *JACM,* Vol. 12, No. 1, 1965, pp. 42–52.

4-24. Griffiths, T. V., and S. R. Petrick, "On the relative efficiencies of context-free grammar recognizers," *CACM,* Vol. 8, No. 5, 1965, pp. 289–300.

4-25. Harrison, M. A., *Introduction to Switching and Automata Theory,* New York: McGraw-Hill, 1965.

4-26. Hopcroft, J. E., and J. D. Ullman, *Formal Languages and Their Relation to Automata,* Reading, Mass.: Addison-Wesley, 1969.

4-27. Ingerman, P. Z., *A Syntax Oriented Translator,* New York: Academic Press, 1966.

4-28. Irons, E. T., "A syntax directed compiler for ALGOL 60," *CACM,* Vol. 4, No. 1, 1961, pp. 51–55.

4-29. Kleen, S. C., "Representation of events in nerve nets," (in) *Automata Studies,* C. E. Shannon and J. McCarthy (eds.), Princeton, N.J.: Princeton University Press, 1956, pp. 3–40.

4-30. Knuth, D. E., "On the translation of languages from left to right," *Information and Control,* Vol. 8, No. 6, 1965, 607–639.

4-31. Knuth, D. E., "Top-down syntax analysis," *Acta Informatica,* Vol. 1, No. 2, 1971, pp. 79–110.

4-32. Knuth, D. E., "Semantics and context-free languages," *Math. Systems Theory,* Vol. 2, No. 2, 1968, pp. 127–146.

4-33. Kuno, S., and A. G. Oettinger, "Multiple path syntactic analyzer," *IFIP Congress 1962,* Popplewell (ed.), Amsterdam: North-Holland, 1962, pp. 306–311.

4-34. McKeeman, W. M., J. H. Hanning, and D. B. Wortman, *A Compiler Generator,* Englewood Cliffs, N.J.: Prentice-Hall, 1970.

4-35. McNaughton, R., and H. Yamada, "Regular expressions and state graphs for automata," *IRE Trans. EC-9,* Vol. 1, 1960, pp. 39–47.

4-36. Parikh, R. J., "On context-free languages," *JACM,* Vol. 13, No. 4, 1966, pp. 570–581.

4-37. Paull, M. C., and S. H. Unger, "Structural equivalence and LL-k grammars," *IEEE Conf. of Record of Ninth Annual Symposium on Switching and Automata Theory,* 1968, pp. 176–186.

4-38. Post, E. L., "Recursive unsolvability of a problem of Thue.," *J. Symb. Logic,* Vol. 12, 1947, pp. 1–11.

4-39. Rabin, M. O., and D. Scott, "Finite automata and their decision problems," *IBM. J. of Research and Development,* Vol. 3, 1959, pp. 114–125.

4-40. Rosenkrantz, D. J., "Matrix equations and normal forms for context-free grammars," *JACM,* Vol. 14, No. 3, 1967, pp. 501–507.

4-41. Rosenkrantz, D. J., and P. M. Lewis, II, "Deterministic left-corner parsing," *IEEE Conf. Record of 11th Annual Symposium on Switching and Automata Theory,* 1970, pp. 139–152.

4-42. Salomaa, A., *Theory of Automata,* New York: Pergamon Press, 1969.

4-43. Samelson, K., and F. L. Bauer, "Sequential formula translation," *CACM,* Vol. 3, No. 2, 1960, pp. 76–83.

4-44. Wirth, N., and H. Weber, "EULER—a generalization of ALGOL and its formal definition," *CACM,* Vol. 9, 1966, pp. 13–23, 89–99.

4-45. Wood, D., "Bibliography 23: formal language theory and automata theory," *Computing Reviews,* Vol. 11, No. 9, 1970, pp. 417–430.

4-46. Younger, D. H., "Recognition and parsing of context free languages in time n^3," *Information and Control,* Vol. 10, No. 2, 1967, pp. 189–208.

5
Sorting

5.1 INTRODUCTION

Many sorting strategies can be expressed most simply as recursive functions because the sorting of a whole collection can often be defined in terms of the same strategy applied to subcollections. There are interesting correspondences between different sorting methods; and the simpler methods appear as components of more complex methods. The strategies are more like variations on a theme rather than distinct programs. The same algorithm will often be expressed in more than one way. The first program to be written operates on a tree-like data structure. This program will then be adapted to operate on a vector, or one-dimensional array, by choosing particular representations of the constructing and selecting functions. Sorting programs are based upon comparisons and simple moving or exchanging operations. The part of the item that determines the ordering is called a *key.* The ordering relation must be applicable to all the objects to be sorted and either it or its converse must hold between any two items. The simplest sorting methods either 1) gradually create a larger sorted collection by adding one key at a time, or 2) remove the smallest key from a collection, leaving the same kind of collection.

Inserting. The function *insert,* defined below, adds a new key k into a list of keys sorted in ascending order by inserting the new key before the first key greater than k.

$$\begin{aligned}
&\textbf{def rec } \textit{insert } k \; x \; = \\
&\qquad \textbf{if } \textit{null } x \\
&\qquad \textbf{then } k:() \\
&\qquad \textbf{else if } k < h \; x \\
&\qquad\qquad \textbf{then } k:x \\
&\qquad\qquad \textbf{else } h \; x : \textit{insert } k \, (t \; x)
\end{aligned}$$

A list can be sorted by repeated insertion of its elements using the function ($\textit{list} 2 \; () $ insert I) or ($\textit{list} 1 \; () $ insert I).

Selecting.   A second way to sort a list is to find the smallest element and remove it. The smallest key can be selected from a nonnull list of keys by the function *select,* defined below, which produces a pair whose first member is the smallest key and whose second item is the remaining list.

$$\begin{aligned}
&\textbf{def rec } \textit{select } x \; = \\
&\qquad \textbf{if } \textit{null}(t \; x) \\
&\qquad \textbf{then } h \; x \\
&\qquad \textbf{else let } y, z \; = \; \textit{select}(t \; x) \\
&\qquad\qquad \textbf{if } h \; x < y \\
&\qquad\qquad \textbf{then } h \; x, \; y:z \\
&\qquad\qquad \textbf{else } y, \; h \; x:z
\end{aligned}$$

The sorted list can be produced by successive selections of the smallest item, as follows

$$\begin{aligned}
&\textbf{def rec } \textit{sort } x \; = \\
&\qquad \textbf{if } \textit{null } x \\
&\qquad \textbf{then } () \\
&\qquad \textbf{else let } y, z \; = \; \textit{select } x \\
&\qquad\qquad y:\textit{sort } z
\end{aligned}$$

Merging.   A third simple method of sorting is by *merging* two sorted lists or strings. The smallest head of the two lists is selected and removed, leaving two sorted lists which then have to be merged in the same way.

$$\begin{aligned}
&\textbf{def rec } \textit{merge } x \; y \; = \\
&\qquad \textbf{if } \textit{null } x \\
&\qquad \textbf{then } y \\
&\qquad \textbf{else if } \textit{null } y
\end{aligned}$$

$$\textbf{then } x$$
$$\textbf{else if } h\,x < h\,y$$
$$\textbf{then } h\,x : merge(t\,x)y$$
$$\textbf{else } h\,y : merge\,x(t\,y)$$

A nonempty list of items can be sorted by merging. Each item is changed into a 1-list and repeatedly merged until one sorted list remains

def rec *mmerge x* $=$

$$\textbf{let rec } mergepairs\,y =$$
$$\textbf{if } null\,y$$
$$\textbf{then } ()$$
$$\textbf{else if } null\,(t\,y)$$
$$\textbf{then } y$$
$$\textbf{else } (merge(h\,y)(h(t\,y)) : mergepairs(t(t\,y)))$$
$$\textbf{if } null\,(t\,x)$$
$$\textbf{then } h\,x$$
$$\textbf{else } mmerge(mergepairs\,x)$$

The number of items transferred from one string to another is the weight of the external nodes of the tree corresponding to the merging pattern. There are, for example, essentially three patterns for merging the five 1-lists *v, w, x, y, z,* namely:

$$merge(merge\,v\,w)(merge(merge\,x\,y)z),$$
$$merge\,v(merge\,w(merge\,x(merge\,y\,z)))$$
$$merge\,v(merge(merge\,w\,x)(merge\,y\,z))$$

The weights of the corresponding trees are 12, 14, and 13, respectively. The most efficient strategy is to use the most balanced tree [5–5].

5.2   SORTING VECTORS

A fourth method of sorting, called *exchanging,* is applicable to a vector of keys. The basic operation is a comparison of the items in two positions, followed by a possible exchange or *swap.* Each exchange transforms the vector to a more sorted arrangement by moving small keys one way and large keys the other. Both the inserting and selecting strategies can be adapted to vectors and expressed in terms of exchanges. Throughout this chapter the vector sorting methods will operate on vector A of length n that is indexed by the integers 1 to n. A list of positions

$$p_1, p_2, p_3, \ldots, p_m$$

of this vector will be called a *chain*. The result of sorting a chain is to rearrange the keys so that $A[p_1], A[p_2], \ldots, A[p_m]$ are in nondecreasing order. The infixed operator $:=:$ will be used to denote an exchange. The *insert* function can be adapted to produce a function called *insertc* which operates on a nonnull chain $(p_1, p_2, p_3, \ldots, p_m)$. It will insert $A[p_1]$ into the sorted list $A[p_2], A[p_3], \ldots, A[p_m]$ by exchanging.

> **def rec** *insertc A x* =
>> **if** *null*$(t\ x)$
>> **then** *exit*
>> **else if** $A[h\ x] < A[h(t\ x)]$
>>> **then** *exit*
>>> **else** $A[h\ x] :=: A[h(t\ x)]$
>>> *insertc A*$(t\ x)$

A chain x can be sorted by inserting items one at a time, as follows

> **def rec** *sort A x* =
>> **if** *null x*
>> **then** *exit*
>> **else** *sort A*$(t\ x)$
>> *insertc A*$(h\ x:t\ x)$

The chain can be specified by a stream; and, when the stream is (1 **step** 1 **until** n), the sorting program takes the simple form below called *straight insertion*. An example of the steps taken when sorting a vector by straight insertion is given in Fig. 5.1.

> **for** $i := n - 1$ **step** -1 **until** 1
>> **do for** $j := i$ **step** 1 **until** $n - 1$
>>> **do if** $A[j] < A[j + 1]$
>>>> **then** *exit*
>>>> **else** $A[j] :=: A[j + 1]$

Sorting by repeated selection of the smallest key can also be adapted to sorting vectors. In this case the position of the smallest is found first; then

```
5   3   1   6   4       2
5   3   1   6       2   4
5   3   1       2   4   6
5   3       1   2   4   6
5       1   2   3   4   6
    1   2   3   4   5   6
```

Fig. 5.1   Straight insertion

the smallest key is exchanged with the first key.

> **def rec** *find* x =
>   **if** *null*$(t\ x)$
>   **then** $h\ x$
>   **else let** $y = find(t\ x)$
>   **if** $A[h\ x] < A[y]$
>   **then** $h\ x$
>   **else** y

The complete sort can be accomplished by repeated selection

> **def rec** *sort* x =
>   **if** *null* x
>   **then** x
>   **else** $A[h\ x] :=: A[find\ x]$
> $h\ x:sort(t\ x)$

In this program it is assumed that it is permissible to exchange an item with itself $(A[x] :=: A[x])$. The chain can be represented by a stream, and the particular chain (1 **step** 1 **until** n) gives rise to the simple program called *linear selection*. An example of the stages in sorting by linear selection is given in Fig. 5.2.

> **for** $i := 1$ **step** 1 **until** $n - 1$
>   **do** $k := i$
>   **for** $j := i + 1$ **step** 1 **until** n
>   **do if** $A[k] > A[j]$
>   **then** $k := j$
> $A[i] :=: A[k]$

$$
\begin{array}{cccccc}
5 & 3 & 1 & 6 & 4 & 2 \\
1 & 3 & 5 & 6 & 4 & 2 \\
1 & 2 & 5 & 6 & 4 & 3 \\
1 & 2 & 3 & 6 & 4 & 5 \\
1 & 2 & 3 & 4 & 6 & 5 \\
1 & 2 & 3 & 4 & 5 & 6
\end{array}
$$

Fig. 5.2   Linear selection

The position of the smallest key found so far is kept in k and after searching the list the smallest key is exchanged with the first key. The same algorithm is then applied to the tail of the list. A list is represented as a pair (i, n) $1 \le i \le n + 1$. The list is null if $i = n + 1$, otherwise the head of the list is i and its tail is $(i + 1, n)$.

Merging.   A vector can be sorted by merging. In this case a second vector P will be used to store the tails of the strings. If $A[i]$ is the head of the list then $P[i]$ contains the position of the next item in the sorted string. The empty list will be represented by zero. The representations of the list functions follow.

null x	$x = 0$
h x	$A[x]$
t x	$P[x]$
prefix x y	$P[x] := y; x$
$()$	0

Using these representations the *merge* function can now be written as follows:

> **def rec** *merge x y =*
> > **if** $x = 0$
> > **then** y
> > **else if** $y = 0$
> > > **then** x
> > > **else if** $A[x] < A[y]$
> > > > **then** $P[x] := merge\ P[x]\ y;\ x$
> > > > **else** $P[y] := merge\ x\ P[y];\ y$

The initial state has $P[i] = 0$, $1 \le i \le n$ to represent each item as a 1-list. The complete sort can be accomplished by splitting the interval $(1, n)$ into two nearly equal segments which are sorted separately and then merged.

The ÷ operator will be used here and throughout this chapter to indicate
the integer part of the result of the division, as in ALGOL 60.

$$sort\ 1\ n$$

> **where rec** *sort x y =*
>> **if** $x = y$
>> **then** y
>> **else let** $c = (x + y) \div 2$
>>> $merge(sort\ x\ c)(sort(c + 1)y)$

The maximum number of comparisons is given by

$$S(n) = S\lceil n/2 \rceil + S\lfloor n/2 \rfloor + n - 1$$
$$S(1) = 0$$

which has the solution

$$S(n) = nk - 2^k + 1\ where\ k = \lceil log_2\ n \rceil$$

Merging streams.   The merging operation can be adapted in another way
so that it operates on and produces *streams* rather than lists.

> **def rec** *merge x y =*
>> **if** *nulls x*
>> **then** *y*
>> **else if** *nulls y*
>>> **then** *x*
>>> **else if** *hs x < hs y*
>>>> **then** *prefixs(hs x)(merge(ts x)y)*
>>>> **else** *prefixs(hs y)(merge x (ts y))*

It is therefore possible to combine a list of streams by merging to produce
a stream for the sorted list. It is more efficient to represent the streams using
a buffer of size 1 because the head of the stream is to be referred to more
than once. Each stream will be represented by a pair whose first is the head
of the sequence and whose second is the old stream representation of the
tail. The new way of representing streams can be summarized as follows:

nulls s	$(h\ s) = end$
nullists	*(end, generate I end)*
hs	*h*
ts(x, s)	*s()*
prefixs x s	$x, \lambda().s$

With this change of representation the *merge* function becomes:

def rec *merge x y =*

> let *u, f* = *x*
> let *v, g* = *y*
> if *u* = *end*
> then *y*
> else if *v* = *end*
>> then *x*
>> else if *u* < *v*
>>> then *u,* λ().*merge*(*f*())*y*
>>> else *v,* λ().*merge x*(*g*())

The function *mmerge* produces a stream-pair whose first is the smallest item, and whose second is a stream which takes the form of a tree or tournament, called a *loser tree*. The application of *mmerge* requires $n - 1$ comparisons. The production of the remaining stream elements requires at most $log_2 n$ comparisons each.

Sorting networks.   The inserting and selecting strategies can also be represented as variations of a fixed pattern of operations, made up of a comparison followed by a possible exchange, called a *comparator*. Each comparison is always made, regardless of the results of previous comparisons. The pattern of comparisons can be written as a network, or circuit diagram, whose elements will be represented in the two ways illustrated in Fig. 5.3. The time sequence is down the page and the smaller key always moves to the left. The networks for the straight insertion and linear selection methods are shown in Figs. 5.4 and 5.5.

A merging network.   Sorting by merging can also be expressed by comparison network. A comparator is both a merging network and a sorting network for two items. Four items are sorted by first producing two strings of length 2 and then merging them according to the pattern illustrated in Fig. 5.6.

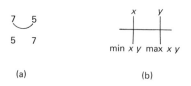

(a)   (b)

Fig. 5.3   Comparison network elements

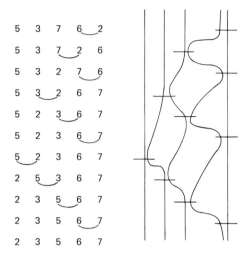

```
5   3   7   6   2
5   3   7   2   6
5   3   2   7   6
5   3   2   6   7
5   2   3   6   7
5   2   3   6   7
5   2   3   6   7
2   5   3   6   7
2   3   5   6   7
2   3   5   6   7
2   3   5   6   7
```

Fig. 5.4   Straight insertion network

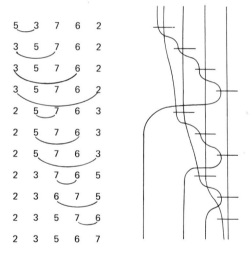

```
5   3   7   6   2
3   5   7   6   2
3   5   7   6   2
3   5   7   6   2
2   5   7   6   3
2   5   7   6   3
2   5   7   6   3
2   3   7   6   5
2   3   6   7   5
2   3   5   7   6
2   3   5   6   7
```

Fig. 5.5   Linear selecting network

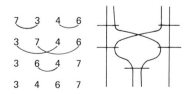

```
7   3   4   6
3   7   4   6
3   6   4   7
3   4   6   7
```

Fig. 5.6   An odd-even merge-sorting network for four items

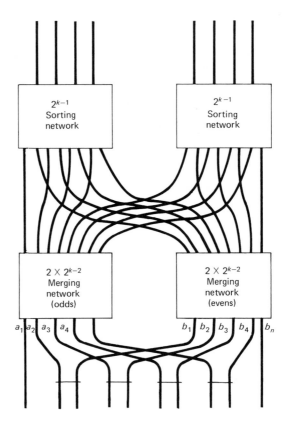

Fig. 5.7 The construction of a merge-sorting network

In general a network for sorting 2^k items is made up of two sorting networks for 2^{k-1} items and one merging network for 2 strings of length 2^{k-1}. A merging network for 2^k items is made up of two 2^{k-1} merging networks plus $2^{k-1} - 1$ comparators. The general method of constructing a 2^k sorting network is shown in Fig. 5.7.

First the odd items in each string are merged, and the even items are merged to produce:

$$a_1\, a_2\, a_3\, a_4 \ldots \text{ and } b_1\, b_2\, b_3\, b_4 \ldots$$

as illustrated in Fig. 5.7. Now a_1 is the smallest, and b_n is the largest, where $n = 2^{k-1}$. Finally, $2^{k-1} - 1$ comparators are used to compare:

$$(a_2, b_1), (a_3, b_2), (a_4, b_3), \ldots, (a_n, b_{n-1}).$$

It is not immediately obvious that such a network sorts numbers properly.

The simplest proof that this method works is by Bouricius' theorem [5–29]. Suppose that an algorithm or machine which is alleged to sort is constructed from comparison elements which compare two numbers. One action is prescribed for greater than, and another action is for less than; but when the two numbers are equal either action can occur in an unpredictable manner. Bouricius' Theorem states that: If this algorithm will sort all files in which only two different numbers (such as $\{0,1\}$) occur as keys, then it can sort any file of numbers.

This theorem can be used to prove that the merging network sorts properly. Suppose the first string contains s zeroes and the second contains t zeroes. Then $a_1, a_2, \ldots$ contains $\lceil s/2 \rceil + \lceil t/2 \rceil$ zeroes and $b_1, b_2, \ldots$ contains $\lfloor s/2 \rfloor + \lfloor t/2 \rfloor$ zeroes. Therefore the number of zeroes in a and b differ by zero, (both even), one (one odd one even), or two (both odd). If they differ by one or zero then

$$a_1\, b_1\, a_2\, b_2\, a_3\, b_3 \ldots$$

are in order. If they differ by 2 then the zeroes and ones must be arranged as follows

$$a\, b\, a\, b\, a\, b\, a\, b\, a\, b\, a \ldots$$
$$0\, 0\, 0\, 0\, 0\, 0\, 0\, 1\, 0\, 1\, 1 \ldots$$

and the final comparisons will sort this arrangement completely.

If $M(n)$ is the number of comparators needed for a merging network, and $S(n)$ is the number needed for a sorting network then

$$S(2^k) = 2\, S(2^{k-1}) + M(2^k)$$
$$M(2^k) = 2\, M(2^{k-1}) + 2^{k-1} - 1$$
$$S(2) = M(2) = 1$$
$$M(2^k) = (k-1)2^{k-1} + 1$$

and it follows that

$$S(2^k) = (k^2 - k + 4)2^{k-2} - 1.$$

The strategy used in the merging network can be expressed in terms of sorting chains. The two chains

$$(1, 3, 5, 7, \ldots, 2q - 1);\ (2, 4, 6, \ldots, 2q)$$

are first sorted, after which the two chains

$$(1, 2, 5, 6, 9, 10, \ldots);\ (3, 4, 7, 8, 11, 12, \ldots)$$

are sorted, and finally the $q - 1$ chains

$$1;\ (2, 3);\ (4, 5);\ (6, 7) \ldots (2q - 2, 2q - 1);\ 2q$$

The chains are sorted using the same method. It has been assumed that
the ordering produced by sorting the second two chains preserves the order
established by sorting the first two chains. This is a special case of a phenomenon which has been investigated by Gale and Karp [5–16], who call
a number of disjoint chains which include all the items to be sorted a *shift*.
They have investigated, among other things, the conditions under which
sorting one shift preserves the ordering established by another. There are
three such conditions.

1. The union of the two shifts can have no directed cycles, so that it is
   not possible to go from any position to itself by following chains composed from the two shifts.
2. Two adjacent positions in a chain of one shift do not belong to the same
   chain of the second shift.
3. The two shifts must *commute,* which is to say that if it is possible to
   get from one position to another by following a chain of the first shift,
   followed by a chain of the second shift; then it is possible to get there
   by following a chain of the second shift, followed by a chain of the first
   shift.

The immediate consequences of this theorem are considered in the following
two-dimensional array. The three conditions are satisfied for the two shifts
made up of the chains of columns and the chains of rows. It follows that
if a two-dimensional array is first sorted by rows and then by columns, it
remains sorted by rows. An example is given in Fig. 5.8. Gale and Karp
have also proved the surprising fact that if the number of rows is greater
than or equal to the number of columns then sorting by the transverse
diagonal (i.e., from southwest to northeast) preserves the sorting by both
rows and columns. The last array in Fig. 5.8 is an example.

7	5	6		5	6	7		1	2	3		1	3	5
3	1	2		1	2	3		3	4	7		2	4	7
3	8	4		3	4	8		5	6	8		3	6	8

by rows   by columns

Fig. 5.8 Column sorting preserves row sorting

Shell sort. A one-dimensional vector can be treated as a p-rowed two
dimensional array whose (i, j) element is indexed by $i + p \times (j - 1)$. If
each row is sorted then the vector will be said to be p-*sorted.* The Shell-sort

program performs p-sorting for a diminishing sequence of integers p, and ends up with a 1-sort, which is a complete sort. The p-sorting program sorts the chains $whiles(\leq n)$ $(generate(+ p)i)$ for $1 \leq i \leq p$ using $insertc$. It is a consequence of the theorem of Gale and Karp that a p-sorted vector remains p-sorted after it has been q-sorted. The p-sorting program can be written as:

> **for** $j := 1$ **step** 1 **until** $n - p$
>> **do for** $i := j$ **step** $- p$ **until** 1
>>> **if** $A[i + p] > A[i]$
>>> **then** $A[i + p] :=: A[i]$
>>> **else** *exit*

Fig. 5.9 is an example of a 7, 3, 1-Shell sort. A variation of Shell sort [5–30] arises from the observation that a vector which is both 3-sorted and 2-sorted does not need the full straight insertion program to 1-sort it. Only adjacent keys need be compared. Furthermore, if $A[i]$ and $A[i + 1]$ are interchanged then it is not necessary to compare $A[i + 1]$ and $A[i + 2]$, because the vector is 2-sorted. The variation uses the sequence of values $2^s 3^t$ for p, and the program can be arranged so that the 2-sorting and 3-sorting use the same technique as 1-sorting. A program based on this idea, however, would 6-sort twice; once as a 2-sort component of a 3-sort, and once as a 3-sort

```
15   13   1   7   5   14   9   4   2   8   11   3   12   16   6

              15   4   6                 4    6   15
              13   2                      2   13
               1   8                      1    8
 7-sort        7  11                      7   11
               5   3         to           3    5
              14  12                     12   14
               9  10                      9   10

i.e., 4   2   1   7   3   12   9   6   13   8   11   5   14   10   15

          4  7  9  8  14          4  7  8  9  14
 3-sort   2  3  6 11  10   to     2  3  6 10  11
          1 12 13  5  15          1  5 12 13  15

i.e., 4   2   1   7   3   5   8   6   12   9   10   13   14   11   15

1-sort    1   2   3   4   5   6   7   8   9   10   11   12   13   14   15
```

Fig. 5.9   Shell-sorting

component of a 2-sort. The 2-sort that follows a 3-sort can be implemented as a 2-sort followed by an interchange program, such as:

$$sort1\ 1\ n$$

where rec $sort\ a\ p\ n\ =$

$$sort3\ a\ p$$
$$sort2\ a\ p$$
$$interchange\ a\ p$$

and $sort3\ a\ p\ =$

if $a + 3p > n$
then $exit$
else $sort\ a\ (3p)$
$$sort\ (a + p)\ (3p)$$
$$sort\ (a + 2p)\ (3p)$$

and $sort2\ a\ p\ =$

if $a + 2p > n$
then $exit$
else $sortm\ a(2p)$
$$sortm(a + p)(2p)$$

and $sortm\ a\ p\ =$

$$sort2\ a\ p$$
$$interchange\ a\ p$$

and $interchange\ a\ p\ =\ intch\ 0$

where rec $intch(q)\ =$
if $a + p + q \leq n$
then $exit$
else if $A[a + q] < A[a + p + q]$
then $A[a + q] :=: A[a + p + q]$
$$intch(q + 2p)$$
else $intch(q + p)$

5.3   BINARY SEARCH TREES

Another way to store the sorted keys is in a *binary search tree*. The keys in a binary search tree are arranged so that the left and right subtrees are also binary search trees. All keys in the left tree are less than the root, and all keys in the right tree are greater. The basic insertion step compares the new key with the root, and inserts it into the left tree if it is greater than the root, and into the right tree if it is less. The shape of the tree depends on the order in which the keys are inserted. Once inserted, a key is never moved again. The following function inserts a key *k* into a tree *x*.

> **def rec** *insertb k x =*
>> **if** *empty x*
>> **then** *cbtree k etree etree*
>> **else if** *k < root x*
>>> **then** *cbtree (root x)(insertb k(left x))(right x)*
>>> **else** *cbtree (root x)(left x)(insertb k(right x))*

The tree can be *flattened* into a sorted list by the function *flatn*.

> **def rec** *flatn t =*
>> **if** *empty t*
>> **then** ()
>> **else** *concat (flatn(left t), u(root t), flatn(right t))*

A list of keys can be inserted one at a time by the function (*list 2 etree insertb I*). The stages in the construction of a binary search tree from the permutation 7 2 9 6 1 8 3 4 5 are shown in Fig. 5.10. One could equally well have

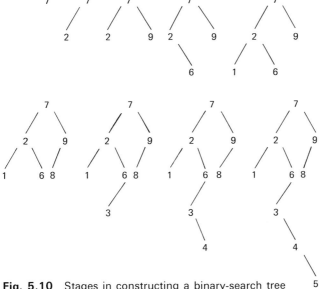

Fig. 5.10   Stages in constructing a binary-search tree

produced a forest by systematically changing the constructing and selecting functions of those for forests according to the scheme:

Binary tree	Forest
etree	()
root x	*root (h x)*
left x	*listing (h x)*
right x	*t x*
cbtree x y z	*(ctree x y): z*

The function for inserting a new item into a forest is therefore:

def rec *insertf k x =*
    if *null x*
    then *(ctree x ()):()*
    else if $k < root (h x)$
        then *ctree (root x)(insertf k (listing(h x))):t x*
        else *ctree (root x)(listing (h x)): insertf k (t x)*

The forest corresponding to 7 2 9 6 1 8 3 4 5 is produced in the stages illustrated in Fig. 5.11. When a key is inserted it is compared with those keys on the path from its position in the final tree to the root; therefore, the expected number of comparisons needed to insert a new key into a binary search tree having *n* keys can be obtained by considering the weights of the trees obtained from all permutations of $\{1, 2, 3, \ldots n\}$. Consider the

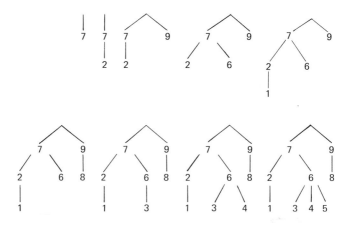

Fig. 5.11 Stages in constructing a forest

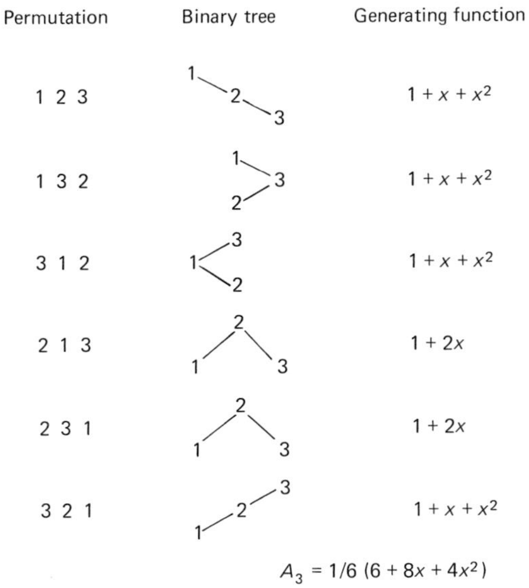

Permutation	Binary tree	Generating function
1 2 3		$1 + x + x^2$
1 3 2		$1 + x + x^2$
3 1 2		$1 + x + x^2$
2 1 3		$1 + 2x$
2 3 1		$1 + 2x$
3 2 1		$1 + x + x^2$

$$A_3 = 1/6 \ (6 + 8x + 4x^2)$$

Fig. 5.12   Binary-search trees

extended binary tree of the form:

- An A-B-binary tree either

   is *atomic* and is a B

   or has a *root* which is an A

   and a *left* and a *right* both of which are A-B-binary trees.

If $A_n(x)$ is the generating function in which the coefficient of x^k is the expected number of internal nodes on level k of a binary search tree with n interior nodes and $B_n(x)$ is the expected number of external nodes on level k of the tree, then $A_n(1) = $ n, the number of internal nodes, and

$$A_n(x) + B_n(x) = 1 + 2xA_n(x).$$

The six binary search trees with three internal nodes are shown in Fig. 5.12, together with the two generating functions. When a new item is added, the tree is extended by changing one B type node into an A type. The probabilities of extending the binary tree at each terminal node are equal. Therefore

$$A_{n+1}(x) = A_n(x) + (1/(n + 1))B_n(x)$$

and if follows that

$$A_{n+1}(x) = (1/(n + 1)) + ((n + 2x)/n + 1)) A_n(x)$$

The expected weight of the interior nodes of a binary search tree is $S_n = A_n^x(1)$. By differentiating $A_{n+1}(x)$ with respect to x and setting $x = 1$, we get

$$S_{n+1} = (2/(n+1))A_n(1) + ((n+2)/(n+1))S_n.$$

But $A_n(1) = n$, the number of interior nodes, therefore:

$$S_{n+1}/(n+2) - S_n/(n+1) = 2n/(n+1)(n+2)$$
$$= 2n(1/(n+1) - 1/(n+2)).$$

Thus, if:

$$H_n = \sum_{k=1}^{n} 1/k,$$
$$S_{n+1}/(n+2) = 2(H_{n+1} - 1) - 2n/(n+2)$$

and

$$S_n = 2(n+1)H_n - 4n.$$

The weight of the external nodes is obtained from B_n, giving

$$B_n^x(x) = 2A_n(x) + (2x-1)A_n^x(x)$$

and

$$B_n^x(1) = 2n + A_n^x(1).$$

Therefore, the expected weight of the external nodes of a binary search tree with n nodes is $S_n + 2n = 2(n+1)H_n - 2n$.

The function *tsearch*, defined below, searches for a key stored in a binary search tree and, if the search is successful, returns the tree whose root is the key, otherwise it returns the empty tree.

```
def rec tsearch k x =
    if empty x
    then x
    else if k = root x
        then x
        else if k > root x
            then tsearch k (left x)
            else tsearch k (right x)
```

The expected number of comparisons to perform an unsuccessful search is the expected number of comparisons needed to insert the $n + 1$st element into an n-element binary search tree. The expected weight of the $n + 1$ external nodes is $2(n+1)H_n - 2n$. Therefore the expected number of comparisons for an unsuccessful search is $(2(n + 1)H_n - 2n)/(n + 1) = 2H_{n+1} - 2$. The expected number of comparisons for a successful search is one more than the expected internal path length of the tree and is therefore $1 + (2(n + 1)H_n - 4n)/n = (2(n + 1)/n)H_n - 3$.

An item k can be deleted from a tree by first finding the tree whose root is k and then removing the root. The technique for replacing a tree by a tree with its root removed is most easily expressed in terms of forests. The root of the head of the list is removed, leaving two forests which can then be concatenated. The result is a forest which when read in postorder produces the items in order. The equivalent operation on binary trees is to replace the rightmost empty tree in the left tree by the right tree.

> **def rec** *replace y x* =
>> **if** *empty x*
>> **then** *y*
>> **else** *ctree*(*root x*)(*left x*)(*replace y*(*right x*))

5.4   BINARY INSERTION

A segment (m, n) of a sorted vector A can be treated as a binary search tree whose root occupies the center position $c = (m + n + 1) \div 2$ and whose left and right trees are the segments $(m, c - 1)$ and $(c + 1, n)$, respectively. The binary insertion program finds the position that a new key should occupy and then moves items to the right of that position to make space for it. The positions that a new key can occupy in a segment (m, n) will be numbered $m, m + 1, m + 2, \ldots, n, n + 1$ and treated as a combination. The combination will be represented by a pair (m, n). An atomic combination has $m = n$. The left of a nonatomic combination is $(m, c - 1)$ and its right is (c, n) where $c = (m + n + 1) \div 2$. The program *posn* for finding the position in a sorted segment (m, n) that a new key k should occupy is defined below:

> **def rec** *posn k*(*m, n*) =
>> **if** $m = n$
>> **then if** $k < A[m]$
>>> **then** *m*
>>> **else** $m + 1$
>> **else let** $c = (m + n + 1) \div 2$
>>> **if** $k > A[c]$
>>> **then** *posn k*(*m, c* − 1)
>>> **else** *posn k*(*c, n*)

A vector can be completely sorted by successive binary insertions as follows:

> **for** $i := 2$ **step** 1 **until** n
>> **do let** $p = posn\ A[i](1, i - 1)$
>> **if** $p \neq i$
>> **then** $t := A[i]$
>>> **for** $r := i$ **step** -1 **until** $p + 1$
>>>> **do** $A[r] := A[r - 1]$
>> $A[p] := t$

The expected number of comparisons using binary insertion is less than the number for straight insertion. To insert a new key into a sorted vector of size j needs at most $\lceil log_2(j + 1)\rceil$ comparisons and the expected number is

$$k + 1 - 2^k/(j + 1)$$

where

$$k = \lceil log_2(j + 1)\rceil$$

This is because the tree underlying the binary insertion strategy is a balanced tree with $(j + 1)$ endpoints representing the positions that a new key can occupy. The number of comparisons needed to find a position is the path length of that position. The expected number of comparisons needed to sort n keys is

$$\sum_{j=1}^{n} (k + 1 - 2^k/j \text{ where } k = \lceil log_2 j\rceil)$$

and the maximum number is

$$\sum_{j=1}^{n} \lceil log_2 j\rceil = nk - 2^k + 1$$

where

$$k = \lfloor log_2(n + 1)\rfloor$$

Binary search. It is possible to determine whether or not a given key belongs to a sorted vector of keys and, if it does, to give its position by a technique, similar to that of binary insertion, in which the key in question is compared with the middle key. If they are equal the search is successful;

otherwise one of the segments to the left or right of the key is subjected
to the same binary search. The program has failed to find a key when the
segment it is to search is found to be empty. The underlying structure
is a binary tree represented by a segment (m, n). The tree is empty if
$m > n$. If it is nonempty, its root occupies position $c = (m + n) \div 2$;
where the $\div$ operation is assumed to produce the integer portion of the
result of the division. The left and right binary trees are represented by $(m,
c - 1)$ and $(c + 1, n)$, respectively. The program for binary searching follows:

$$
\begin{aligned}
&\textbf{def rec } bsearch\ k\,(m,\ n)\ = \\
&\quad \textbf{if } m > n \\
&\quad \textbf{then } fail \\
&\quad \textbf{else} \\
&\qquad \textbf{let } c = (m + n) \div 2 \\
&\qquad \textbf{if } k = A[c] \\
&\qquad \textbf{then } c \\
&\qquad \textbf{else } bsearch\ k\ (\textbf{if } k < A[c] \\
&\qquad\qquad\qquad\qquad\quad \textbf{then } (m,\ c - 1) \\
&\qquad\qquad\qquad\qquad\quad \textbf{else } (c + 1,\ n))
\end{aligned}
$$

The expected number of equality tests needed to conduct a successful
search for a key which is in the vector is one more than the average path
length of the internal nodes of the tree. The total internal path length is

$$
\sum_{j=1}^{n} \lfloor log_2 j \rfloor = (n + 1)(k + 1) - 2^k - 2n
$$

where

$$
k = \lceil log_2(n + 1) \rceil
$$

5.5 QUICKSORT

There is another algorithm for constructing binary search trees which not
only produces the same binary tree but also performs the same comparisons
in a different sequence. In this method the first number is made the root,
and the remaining items are partitioned into two lists, one containing those
items less than the first and the other containing those which are greater.
The same algorithm is then applied to each partition to form the left and

right subtrees. The function is:

def rec *qs x =*
    if *null x*
    then *etree*
    else let $d = h\,x$
        let *y, z = partition d(t x)*
        cbtree d(qs y)(qs z)
where rec
    partition d x =
        if *null x*
        then (), ()
        else let *y, z = partition d(t x)*
            if $h\,x < d$
            then $h\,x\!:\!y, z$
            else $y, h\,x\!:\!z.$

The tree can be flattened at the same time by using the function *qs*1, defined below, instead of *qs*.

def rec *qs*1 *x =*
    if *null x*
    then ()
    else let $d = h\,x$
        let *y, z = partition d(t x)*
        *concat (qs*1 *y, u d, qs*1 *z)*

This method can be adapted to sort a vector **A**[1], **A**[2], ..., **A**[n] by exchanging, and the resulting program is known as *quicksort.* The keys **A**[2], **A**[3], ..., **A**[n] are partitioned into those less than **A**[1] and those greater than **A**[1]. The partition produces a vector in which all keys less than **A**[1] lie to its left, and all those greater lie to its right. This is done by finding the first key greater than **A**[1] reading from left to right, and the first key less than **A**[1] reading from right to left, and then exchanging these two keys. The rightgoing and leftgoing scans alternate. First **A**[1] is removed leaving a space. Then the first key smaller than **A**[1] reading from the right is found and placed in the open space, vacating its space which will in turn be filled by the first key greater than **A**[1] reading from the left, also leaving a space, ..., etc. The scanning continues until the two pointers meet. At this stage, both pointers address a space which separates the keys. The key A[1] is then placed in this space and the two segments to its left and right are then sorted using the same method. The following *quicksort* program sorts the segment (m, n) of a vector **A.**

def rec *quicksort*(m, n) =

    if $m \geq n$

    then *exit*   **where rec**

    else let $i = $ *partition* (m, n) *partition*(m, n) =

            $A[i] := A[m]$ **let** $d = A[m]$

            quicksort $(m, i - 1)$ **let** $E = $ **JI**

            quicksort $(i + 1, n)$ *down*(m, n)

---

where rec   **and**

    down (i, j) =   *up* (i, j) =

       let $k = $ *findb*(i, j) **let** $k = $ *findf*(i, j)

           $A[i] := A[k]$ $A[j] := A[k]$

           up $(i + 1, k)$ *down* $(k, j - 1)$

---

where rec *findb*(i, j) =   **and rec** *findf*(i, j) =

    if $i = j$ **if** $i = j$

    then $E\ i$ **then** $E\ i$

    else if $A[j] < d$ **else if** $A[i] > d$

        then j **then** i

        else *findb*$(i, j - 1)$ **else** *findf*$(i + 1, j)$

The function *findb* finds the position of the first item less than d reading right to left from j; *findf* finds the first item greater than d reading right to left from i. If the pointers meet, then an exit is made from the *partition* function, delivering the separating position as a result. The mutually recursive functions *down* and *up* are alternately applied. The expected number of comparisons to sort n items by *quicksort* is the same as that for constructing binary search trees with n nodes—$2(n + 1)H_n - 4n$. An example of the steps taken during a partition operation is given in Fig. 5.13.

    The kth smallest item in a binary search tree can be obtained by attaching one plus the number of nodes in the left subtree to each root. This

5	3	1	6	2	4
□	3	1	6	2	4
6	3	1	□	2	4
6	□	1	3	2	4
6	5	1	3	2	4

Fig. 5.13 Steps in partitioning

number is called the *size* of the tree. The algorithm for obtaining the kth item is defined below.

def rec *th k x =*

  if *empty x*
  then *fail*
  else if $k = size\ x$
   then *root x*
   else if $k > size\ x$
    then *th* $(k - size\ x)(right\ x)$
    else *th k* (*left x*)

The quicksort technique can be adapted to finding the kth smallest item in an unsorted vector. First, the partition function is applied, resulting in the position i that the first item is to occupy. If $i = k$, then $\mathbf{A}[k]$ is the kth smallest item; if $k > i$, then the $k - i^{th}$ smallest item is found in the right segment $(i + 1, n)$; otherwise the kth smallest item is found in the left segment $(1, i - 1)$.

def rec *th k*(*m, n*) =

  if $m \geq n$
  then *fail*
  else let $i = partition\ (m, n)$
   $A[i] := A[m]$
   if $k = i$
   then $A[i]$
   else if $k > i$
    then $th(k - i)(i + 1, n)$
    else *th k*(*m, i* − 1)

5.6   RADIX EXCHANGE AND TRIE STORAGE

This method is similar to quicksort in scanning the list of keys from both ends towards the middle. In this case, however, each phase of key separation is determined by inspecting the most-significant binary digit of the key. The scan from the left finds the first 0, the scan from the right finds the first 1, and then they are exchanged. The same process is then carried out on the second most significant digit, etc. After the separation has been made on the least-significant digit, the keys are in order. An example of radix exchange sorting is given in Fig. 5.14. Again there is an associated method of tree storage. In this case, the set of keys is stored as the set of root-leaf paths of the tree, so that common initial segments are only stored once. The set of bytes in the example would, when stored in this way, give rise to the tree in Fig. 5.15.

12	1100	0001	0001	0001	0001	1
13	1101	0100	0011	0011	0011	3
6	0110	0110	0110	0101	0100	4
15	1111	0011	0100	0100	0101	5
10	1010	0101	0101	0110	0110	6
5	0101	1010	1010	1000	1000	8
14	1110	1110	1000	1010	1010	10
8	1000	1000	1110	1100	1100	12
3	0011	1111	1111	1101	1101	13
4	0100	1101	1101	1111	1110	14
1	0001	1100	1100	1110	1111	15

Fig. 5.14   An example of radix exchange sorting

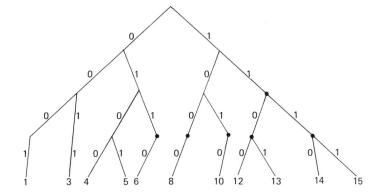

Fig. 5.15   Trie storage and radix exchange

This method of storing a set of keys is called *trie storage*. In general, this tree has the structure of a forest. The words stored in such a forest can be of varying length. Since one word can be a prefix of another it is necessary to store an end indication in the trie. The methods of quicksort and radix exchange are brought into correspondence by supposing that in the radix-exchange trees the decision on the most significant 0 or 1 is in fact made by a comparison with 2^n for suitable n; i.e., 0 corresponds to $<$, 1 to $\geq$. These are the imaginary occupants of internal nodes. The number of comparisons or bit inspections is the sum of the path lengths of terminal nodes of the tree.

The function for adding a new list to a forest is:

def rec *addtree k x* =

    if *null k*

    then *x*

    else if *null x*

        then *(ctree(h k)(addtree(t k)x)):()*

        else if *(h k)* = *root (h x)*

            then *ctree (root(h x))*

                    (addtree (t k)(listing x):t x)

            else *h x:addtree k(t x)*

The function for searching a tree for a list of items is called *searchtree*.

def rec *searchtree k x* =

    if *null k*

    then true

    else if *null x*

        then false

        else if *(h k)* = *root (h x)*

            then *searchtree (t k)(listing (h x))*

            else *searchtree k(t x)*

5.7   TREE INSERTION AND PROMOTION

There is an algorithm which is closely related to sorting with a binary search tree. The keys are stored in a binary tree, but this time they are ordered in the parent-offspring direction. In the case of the associated forest a root is less than any item in its listing, and the items in a listing are in ascending order. A forest is constructed by adding items one at a time. Each entering item is compared with the roots of the top-level trees in the existing forest, and a new tree is constructed which has the entering item as its root and all the top-level trees whose root numbers are greater than itself as its immediate subtrees. This new tree is then added to the remaining trees to form the new forest. A root number is, therefore, always less than the root numbers of its subtrees, and always greater than the root number of its left neighbor tree. The numbers at the roots of the trees in each forest are thus in ascending order.

The function for inserting a tree x into a forest y is called *insert* 1. The stages in the construction of a forest from the permutation 5 2 7 8 9 4 1 6 3 are given in Fig. 5.16; items are inserted one at a time from the right.

> **def rec** *insert* 1 *x y* =
>> **if** *null y*
>> **then** x :()
>> **else if** *root x* > *root* (*h y*)
>>> **then** x : y
>>> **else** *insert* 1(*ctree*(*root x*)(*h y* : *listing x*))(*t y*)

Note first that the correspondence between permutations and forests labeled in this way (i.e., with an ascending order both in the parent-offspring direction and from left to right within each listing) is unique. This is because the permutation that produced a forest can be recovered from the forest by selecting from each listing in order and from a listing immediately before its root. This method of reading is called *postorder*. The operations that construct forests do not change the sequence obtained by this method of reading.

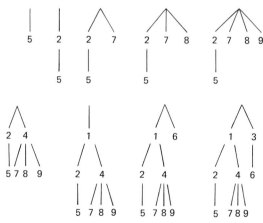

Fig. 5.16 Constructing a forest using insert1

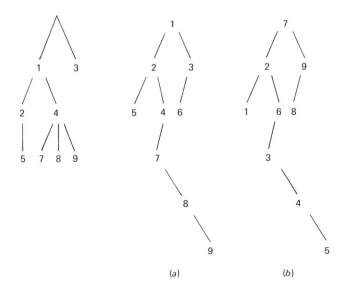

Fig. 5.17 A correspondence between the forests and binary search trees

For purposes of analysis it is convenient to consider the binary trees that underlie the forests which have been produced. The forest produced in Fig. 5.9 together with its corresponding binary tree are shown in Fig. 5.17. The numbers in the binary tree are in ascending order only down the tree.

Fig. 5.17 also contains a copy of the binary search tree produced in Fig. 5.10. It should be noted that the two binary trees in Figs. 5.17 are the same shape. Tree (a) is the binary tree which underlies the forest produced by repeatedly inserting items of the permutation 5 2 7 8 9 4 1 6 3. Tree (b) is the binary search tree of the permutation 7 2 9 6 1 8 3 4 5, which is the inverse of 5 2 7 8 9 4 1 6 3. It is no accident that these two trees have the same shape. The binary tree underlying a forest produced by repeatedly applying *insert* 1 to a permutation p, generally does have the same shape as the binary search tree of the inverse of p. Moreover, if

$$p_1 p_2 p_3 \cdots p_n$$

is the permutation which produced one of these trees then p_i occupies the same position in it as i does in that produced from the inverse permutation. This is most easily seen by considering another way of producing the underlying binary tree for the forest produced by *insert* 1. The same tree could have been produced by selecting the smallest item in the permutation to serve as the root, and then treating the preceding items in the permutation similarly to produce the left subtree, and the following items to form the right subtree.

If $p_k = 1$ is the smallest item in the permutation, $p_1 \, p_2 \, p_3 \cdots p_n$, it will occupy the root of the complete tree. Its first item is the root of the binary search tree of its inverse, i.e., k. Thus the theorem is true for the roots of complete trees. The partition into

$$p_1 \, p_2 \, p_3 \cdots p_{k-1}$$

and

$$p_{k+1} \, p_{k+2} \cdots p_n$$

splits the items into those which will be placed in the left and right subtrees. This partition corresponds to the partition of the inverse into those items less than, and those items greater than k in the inverse permutation. These items will be placed in the left and right subtrees of the binary search tree of the inverse. It is again true for each partition that if p_j is the smallest item in it then j is the first item in the corresponding partition of the inverse. It follows from this correspondence that if labels are ignored, the set of binary trees corresponding to the forests produced by the *insert* 1 algorithm is the set of binary search trees of permutations of $\{1, 2, 3, \ldots, n\}$. From the alternative description of the *insert* 1 algorithm it is easy to see that if a permutation p gives rise to a tree, then the reverse of p gives rise to the reversal of the binary tree (i.e., that formed by interchanging all right and left subtrees).

The number of comparisons required by *insert* 1 to construct forests from permutations of $\{1, 2, 3, 4, \ldots, n\}$ will be considered next. It will be found that the number of comparisons depends only on the shape of the forest produced and not upon how it was labeled. Both the construction of the forest and the subsequent selection of the items therefore depend only on comparisons between the items, and consequently only on their relative sizes. One is therefore at liberty to consider the permutations of the first n natural numbers as the information that is to be sorted. When the *insert* 1 algorithm is used, a number at the root of a tree must, in order to become a root, be compared with the roots of its immediate descendants. When a number is at the top level, it is also compared with certain entering items. After comparison, the numbers that follow it in the permutation make up the leftmost edge of the tree which is its right neighbor. If we consider the corresponding binary tree we see that the numbers that are compared with the root lie on the rightmost edge of the left subtree, and the leftmost edge of the right subtree. It is possible to deduce the comparisons that have been made by looking at the shape of the tree produced. The total comparisons required to construct a tree are therefore the number needed to construct the left and right subtrees plus the number of comparisons made with the root. Consequently, the average number of nodes along a left or right edge of a binary tree must be determined.

The forests produced by *insert* 1 have the property that the numbers at the roots of the top level trees are those numbers in the permutation that have no smaller number following them. Any smaller number following would have been placed above it. In the permutation 5 2 7 8 9 4 1 6 3 the numbers 1 and 3 have this property, are at the roots of the top-level trees of the forest, and occupy the rightmost edge of the corresponding binary tree. These numbers will be called *right-to-left local minima* or *minima*. There is a correspondence between two sets of permutations which can be used to show that the distribution of the minima is the same as that of the cycles of permutations. The permutation that corresponds to 5 2 7 8 9 4 1 6 3 is found by placing brackets around each segment which ends with a minimum to form the cycle representation of another permutation, namely: (5 2 7 8 9 4 1)(6 3). The reverse transformation is made by taking the cycle representation, rotating each cycle so that the smallest item appears last in its cycle, then sorting the cycles so that their last elements are in ascending order, and finally removing the brackets. The probability that a permutation of the first n numbers has k minima is the same as the probability of k cycles occurring, and is the coefficient of x^k in $x(x + 1)$ $(x + 2) \ldots (x + n - 1)/(n!)$. The expected number of minima is therefore $H_n = 1 + 1/2 + 1/3 + \ldots 1/n$. This is also therefore the expected number of nodes on the leftmost (or rightmost) edge of the binary trees produced by either method.

If C_n is the expected number of comparisons required to construct a forest with n nodes, then

$$C_1 = 0, C_2 = 1, \text{and} (n + 1)C_{n+1} = 2 \sum_{k=1}^{n} (C_k + H_k),$$

since each division $(k, n - k)$ $k = 1, 2, 3, \ldots, n$ of the sizes of subtrees of an $(n + 1)$ binary tree is equally likely. Therefore

$$C_n = 2(n - H_n)$$

is the expected number of comparisons needed to construct a forest containing n nodes using *insert* 1.

The function *insert* 1 constructs a forest which is the reverse of the forest required for selecting in the second stage. One could either reverse the top level of the forest or use a representation of the top-level list in which access to both ends is possible. If one now assumes that this reversal has been carried out, then the smallest item lies at the root of the first tree. This can be removed, leaving two forests which now have to be merged to form one forest with the same ordering property. There are several ways of merging two forests. The first is similar to that used in tree- or heap-selection methods.

In the first method the roots of the two heads of the forests are compared and the smaller is promoted, leaving two forests that have to be merged in the same way. The merging continues down until one of the forests is empty. The algorithm for merging two forests in this way is given below.

def rec *merge*1 *x y* =
    if *null x*
    then *y*
    else if *null y*
        then *x*
        else
        if *root*(*h x*) < *root*(*h y*)
        then (*ctree*(*root*(*h x*))(*merge*1(*listing*(*h x*))(*t x*))):*y*
        else (*ctree*(*root*(*h y*))*x*):*merge*1(*listing*(*h y*))(*t y*)

A second method of merging is the same as that used in conventional merge sorting with the difference that the items being merged are trees.

def rec *merge*2 *x y* =
    if *null x*
    then *y*
    else if *null y*
        then *x*
        else if *root*(*h x*) < *root*(*h y*)
            then *h x*:*merge*2(*t x*)*y*
            else *h y*:*merge*2 *x*(*t y*)

Another way to produce the final sorted list is to first produce a sorted list from each forest and then merge these two lists in a conventional way. The numbers that are compared just before they pass through a root are the same in both this method and the *merge*1 algorithm. In fact the *merge*1 algorithm is the stream representation of this list algorithm. The total number of comparisons needed by *merge*1 is the number required to sort the two forests plus the number needed to merge the two sorted lists produced.

The expected number of comparisons needed to merge a list of length k with a list of length $n - k$ can be found by considering the possible ways in which the final list was created from the two lists. The items from the k-list will be called a's and those from the other b's. The total number of arrangements of a's and b's in the final list is $\binom{n}{k}$. The two lists are merged requiring one comparison per item until one of the lists is exhausted, after which no further comparisons are required. One is therefore concerned with the number of arrangements of a's and b's that end with a sequence of a's or b's. The number that end with at least t a's is $\binom{n-t}{k-t}$, and the number that end with at least t b's is $\binom{n-t}{n-k-t}$. The expected number of comparisons to merge two strings of lengths k and $n - k$ is therefore

$$n - (1/\binom{n}{k}) \sum_{t=1}^{n} (\tbinom{n-t}{k-t}) + (\tbinom{n-t}{n-k-t}) = n - (n - k)/(k + 1) - k/(n - k + 1)$$

If M_n is the expected number of comparisons needed to sort a forest with n nodes using *merge*1, then

$$(n + 1)M_{n+1} = \sum_{k=0}^{n} (M_k + M_{n-k} + n - (n - k)/(k + 1)$$
$$- k/(n - k + 1))$$

$$= 2 \sum_{k=0}^{n} M_k + (n + 1)(n + 2 - 2H_n)$$

Therefore $M_n = 2(n + 2)H_n - 6n$.

If the expected number of comparisons needed to construct the tree is added to this, one finds that

$$C_n + M_n = 2(n + 2)H_n - 6n + 2(n - H_n)$$
$$= 2(n + 1)H_n - 4n$$

which surprisingly is the average sum of the path lengths of the set of binary search trees, and is also the expected number of comparisons needed to construct a binary search tree.

The forests or binary trees produced at the intermediate stage have a number of interesting properties. The average sum of the path lengths of the forests is $(n + 1)(H_{n+1} - 1)$ or ΣH_k, and the expected depth of the number r in the forests produced is H_r. The expected depth of the number r in the associated binary trees is $2H_r - 2$.

In order to examine the sum of the path lengths or *weight* of the forests, suppose that $T_n(y)$ is a generating function in which the coefficient of y^k is the probability that a forest of size n has weight k. As has been shown above the sizes of the top level trees correspond to the cycle structure of the permutations. For example, since the cycle index for permutations of $\{1, 2, 3\}$ is

$$h_3 = (1/6)(s_1{}^3 + 3s_1 s_2 + 2s_3),$$

it signifies that of the 6 possible forests produced, there will be one with trees of sizes (1, 1, 1), three with trees of sizes (2, 1), and two with one tree, each of size (3). Each tree of size k has a root, and a forest produced in the same way from permutations of the $(k - 1)$ other items of the cycle. The weight of a k-tree that has been shifted down one level is found by adding 1 to each path length, or k to the whole weight. The generating function for a set of k-trees which have all been shifted down one level is found by multiplying its generating function by y^k. It follows that the recurrence relation for the generating function $T_n(y)$ can be written down immediately from the cycle index h_n. So for example

$$T_0 = 1$$
$$T_1 = y$$
$$T_2 = (1/2)((yT_0)^2 + y^2T_1)$$
$$T_3 = (1/6)((yT_0)^3 + 3(yT_0)(y^2T_1) + 2(y^3T_2))$$

In general the recurrence relation for $T_n(y)$ is found by substituting y^kT_{k-1} for s_k in the cycle index h_n. But the cycle indices are given by

$$1 + h_1x + h_2x^2 + h_3x^3 + \ldots = \exp(s_1 + s_2x^2/2 + s_3x^3/3 + \ldots$$

It follows that if

$$T(x, y) = T_0 + T_1x + T_2x^2 + T_3x^3 + \ldots$$

then

$$T(x, y) = \exp(T_0xy + T_1(xy)^2/2 + T_2(xy)^3/3 + \ldots)$$

If this is differentiated with respect to x then

$$T^x(x, y) = yT(xy, y).T(x, y)$$

and by equating coefficients of x^n

$$(n + 1)T_{n+1}(y) = y. \sum_{k=0}^{n} y^k T_k(y)T_{n-k}(y)$$

If this is now differentiated with respect to y, and y is set equal to 1 the relation for the expected weights of the forests is obtained, namely:

$$(n + 1) W_{n+1} = (n + 1)(n + 2)/2 + 2 \sum_{k=0}^{n} W_k$$

which leads to

$$W_n = (n + 1)(H_{n+1} - 1) = \sum_{k=1}^{n} H_k$$

The expected depths of numbers in the forests that correspond to the binary search trees are considered next. The forest produced from the permutation p with 1 deleted is obtained by deleting 1 and its branch from the forest produced from p. It follows that the expected depth of the number r in an n-forest is the expected depth of $r - 1$ in an $n - 1$ forest for $r > 1$, and therefore is the expected depth of 1 in an $n - r + 1$-forest. The expected depth of the number r in an n-forest that corresponds to a binary search tree is, therefore, H_{n-r+1}.

Let the expected depth of r in a binary search tree with n nodes be $B_{n,r}$. Since r cannot occur in the left k-tree when $k < r$, nor in the right n-k-tree when $k > r - 1$, and occurs at the root, with depth zero, when $k = r - 1$; it follows that

$$nB_{n+1, r} = n + \sum_{k=0}^{r-2} B_{n-k, r-1-k} + \sum_{k=r}^{n} B_{k, r}$$

which surprisingly can be reduced to

$$B_{n+1, r+1} - B_{n, r} = B_{n, r+1} - B_{n-1, r}$$
$$= B_{r+1, r+1} - B_{r, r}$$

But the expected depth of r in an r-tree is the expected depth of 1 in an r-tree. This is the expected number of nodes on the leftmost edge minus one (because the root is at level zero) and is equal to $H_r - 1$. So

$$B_{n+1, r+1} - B_{n, r} = (H_{r+1} - 1) - (H_r - 1) = 1/(r + 1)$$

and so

$$B_{n, r} = H_{n-r+1} + H_r - 2$$

5.8   HEAP SORT

It is possible to store a binary tree in a segment of a vector (i, n) using the convention that the root is $A[i]$ and its subtrees are defined by $(2i, n)$ and $(2i + 1, n)$. The binary tree (i, n) is empty if $i > n$, otherwise its root is $A[i]$, its left is $(2i, n)$, and its right is $(2i + 1, n)$. An arrangement in which each root is larger than any item in its subtrees is called a *heap*. A heap is an intermediate stage in sorting a vector in its own area. There are two ways to construct a heap and two methods of sorting a heap. A new item can be added to a heap $(1, n)$ to produce the heap $(1, n + 1)$ by placing the new key in $A[n + 1]$ and inserting it in the path from leaf to root. One method of constructing a heap (Fig. 5.18) is to start with a heap of size

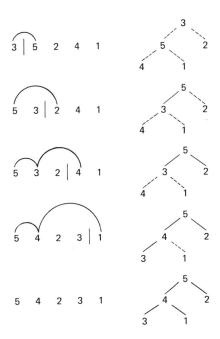

Fig. 5.18   Constructing a heap by inserting through the leaves

1 in $A[1]$ and insert the remainder one at a time. The function for inserting a new item at position p performs a chain insertion using the chain p, $p \div 2$, $(p \div 2) \div 2$, ..., 1. In other words the chain from p to the root is represented by the stream

$$\textbf{def } upchain \; p \; = \; whiles \; (\geq 1)(generate \; (\div 2) \; p)$$

The items in the positions 2, 3, ..., n are inserted in that order using the function *inheap* defined as follows

$$\textbf{def } inheap \; A \; i \; = \; insertc \; A \; (upchain \; i)$$
$$\textbf{for } j := 2 \textbf{ step } 1 \textbf{ until } n$$
$$\textbf{do } inheap \; A \; j$$

Another way to construct a heap is to build it up from two heaps and a new key. The new key is inserted down the tree from root to leaf. It is inserted, using *insertc* with reversed ordering, into the chain formed by choosing the larger key at each bifurcation.

The function for obtaining the chain formed by selecting the largest root at each branching from a binary tree is

> **def rec** *maxpath x* =
>> **if** *empty x*
>> **then** ()
>> **else** *root x* :[**if** *empty*(*left x*)
>>> **then** *maxpath*(*right x*)
>>> **else if** *empty* (*right x*)
>>>> **then** *maxpath*(*left x*)
>>>> **else if** *root*(*left x*) > *root*(*right x*)
>>>>> **then** *maxpath*(*left x*)
>>>>> **else** *maxpath*(*right x*)]

In a heap, however, if the left subtree is empty, then $2i > n$, and the right subtree must also be empty. The only occasion when both subtrees are not empty or nonempty is when $2i = n$ and $2i + 1 > n$. The *maxpath* function can be converted to operate on a heap and produce a stream of positions as follows.

> **def rec** *maxpath*(p, n) =
>> **if** $p > n$
>> **then** *nullists*
>> **else** $\lambda().p,$
>>> **let** $j = 2p$
>>> **if** $j > n$
>>> **then** *nullists*
>>> **else if** $j = n$
>>>> **then** *maxpath*(j, n)
>>>> **else if** $A[j] > A[j + 1]$
>>>>> **then** *maxpath*(j, n)
>>>>> **else** *maxpath*($j + 1, n$)

A heap can be constructed by applying the function *sift* (see Fig. 5.19), defined below:

> **def** *sift i n* = *insertc A* (*maxpath* (*i, n*))

to the nodes that are not endpoints, building up heaps from the bottom by combining two heaps and a key.

> **for** $i := n \div 2$ **step** (-1) **until** 1
>> **do** *sift i n*

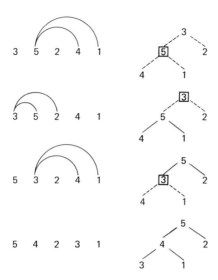

Fig. 5.19 Constructing a heap by inserting through the root

The heap can be completely sorted by using *sift* on decreasing heaps (see Fig. 5.20) as follows

for $i := n$ **step** -1 **until** 2

    do $A[1] :=: A[i]$

        sift $1\,(i-1)$

The exchange puts the root key into its final position, and the key which occupied the final position is then inserted through the root. Each iteration reduces the size of the remaining heap by one.

An alternative method of sorting a heap is to remove the smallest key from the root and repeatedly promote the smallest key in the subtrees until a space is left at an endpoint of the tree. Then the key that occupies the position in which the smallest key is to be placed is inserted through that endpoint using *inheap*. Finally, the smallest key is placed in its final position, reducing by one the size of the heap that remains to be sorted. The sifting operation can be expressed as merely shifting the items in the chain (*max-path* $(1, n)$) up one place as follows:

    def rec *shift s* $=$

                let $x, y = s()$

                if *nulls y*

                then x

                else $A[x] := A[hs\ y]$

                    shift y

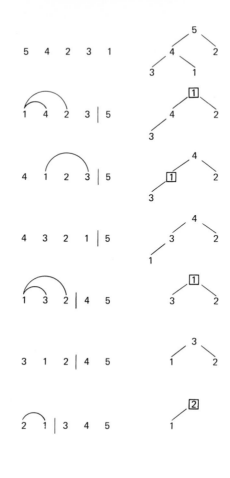

Fig. 5.20 Sorting a heap by inserting through the root

The result is the position of the endpoint. The program for sorting from a heap (see Fig. 5.21) to an ordered vector is

$$\textbf{for } i := n \textbf{ step} - 1 \textbf{ until } 2$$
$$\textbf{do } r := A[1]$$
$$\textbf{let } k = shift\ i$$
$$\textbf{if } k \neq i$$
$$\textbf{then } A[k] := A[i]$$
$$inheap\ A\ k$$
$$A[i] := r$$

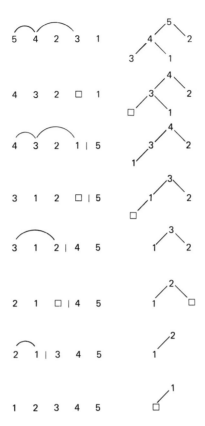

Fig. 5.21　Sorting a heap by inserting through the leaves

5.9　TAPE SORTING

Merging.　Tape-merging programs can be constructed from a merge operation which combines k strings to produce one string whose length is the sum of the lengths of the original strings. In this operation the k original strings and the resulting string must lie on $(k + 1)$ distinct tapes. The complete merging operation starts with an initial set of records on one tape. These are combined into strings as long as possible by internal sorting and distributed on the available tapes. Any merging discipline that continues to combine strings will eventually terminate with one string containing all the records. The problems of tape sorting are those of designing a method which reduces the number of strings as much as possible at each stage and avoids the nonproductive movement of tapes. The strings will always be written forward on the tapes, but can be read either forward or backward. It is unusual to combine forward and backward reading in the same merge program.

The pattern of merging can be represented by a tree in which each node represents the basic operation of a merge of k strings to form a single string. The edges at the nodes will be labeled with a tape letter, and these labels must be distinct. The nodes of the tree will be labeled with distinct integers called *step numbers* to give the order in which the merging steps are carried out. The initial distributions of the strings will usually be omitted from the tree. The end points will all be labeled with zeros which represent the state just after the strings have been distributed. Since strings cannot be merged before they are created, the step number at a node must be greater than the step number at any node in its subtrees.

Another way to specify a merging pattern is by using a table containing entries which either specify the number, and perhaps the length, of strings on each tape, or specify steps which change the state of the tapes. The first entry in this table is the initial state of the tapes. The table has one entry for each merging step. The entry will have -1 for each string read from a tape in that step, $+1$ for each string written, and 0 if the tape does not participate in the step. These step entries are subject to the conditions that it is not possible to read from an empty tape, so that the cumulative sum for each tape cannot be negative. The last entry in the table is the final state. For a complete sort, it should have a 1 representing one string on one tape and a 0 on all the other tapes.

Balanced merging.   The simplest tape-merging method is called *balanced merging*. In a balanced merge the set of tapes available for merging is divided into two disjoint sets. If there are $2k$ tapes they are divided into two sets of size k. If there are $(2k + 1)$ tapes they are divided into two sets, one of size k and the other of size $k + 1$. The strings are first distributed evenly on members of one set, and the merge proceeds by merging strings from one set and distributing them evenly on the other. The transfer of all the strings from one set to the other is called a *pass*. All the items are moved from one set of tapes to the other in a pass, and the time taken for this method is determined by the number of passes needed to sort all the records. The total number of string transfers is the product of the number of passes and the total number of strings. Examples of the table and tree for the balanced sort of nine strings using four tapes are given in Fig. 5.22. The number of passes needed to sort N items when $2k$ tapes are available is the number of levels of the tree, which is $\lceil \log_k N \rceil$. In the example there are 9 strings and $\lceil \log_2 9 \rceil = 4$ passes. Therefore 36 string transfers are made, plus an additional 9 for initial distribution.

The *effective power* of a merge is $S^{(T(S))/S}$, where S is the number of strings, and $T(S)$ is the number of string transfers. In other words, if the number of string transfers is $S \log_p S$, then p is the effective power of the

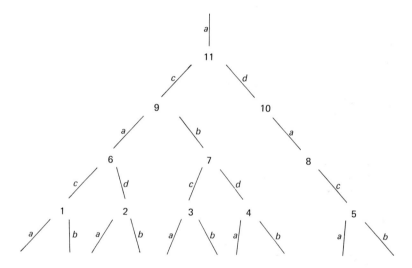

	a	b	c	d
	5	4	0	0
1	-1	-1	+1	0
2	-1	-1	0	+1
3	-1	-1	+1	0
4	-1	-1	0	+1
5	-1	0	+1	0
6	+1	0	-1	-1
7	0	+1	-1	-1
8	-1	-1	+1	0
9	-1	-1	+1	0
10	-1	0	0	+1
11	+1	0	-1	-1
	1	0	0	0

Fig. 5.22 Balanced merging

merge. It is the factor by which the number of strings is reduced when a full pass over all the strings is made. For a balanced merge with $2p$ tapes the effective power is p; and with $2p + 1$ tapes, it is $\sqrt{p(p + 1)}$.

Trees and merging strategies. Both read-forward and read-backward merging will be considered. In the case of read-forward merging, the tape is used as a buffer, or queue. It is written, rewound, and read in the same order as that in which it was written. The read-backward tape is used as a pushdown list. The first string read from a tape is the last string that was written on the tape. Reading backward has the additional complication that a string written in ascending order is read in descending order, and vice versa.

The same table can be used to describe both read-forward and read-backward merging. The rules for constructing read-forward and read-backward merging trees, however, differ. Suppose that one edge of a tree is indicated by the pair of step numbers at its end points in ascending order. An edge (r, s) represents a string which is written on the tape at step r and is read from the tape at step s. When reading forward, the first string written is the first string to be read. So for any two strings (r, s) and (t, u) on the same tape, if $r < t$ then $s < u$. When reading backward, the first string written is the last string to be read; thus if $r < t < s$, then $u < s$. In other words if r, s, t, and u are arranged in ascending order then the lines joining r to s and t to u have no crossovers.

The rules for constructing merging trees can be summarized as follows. A tree represents a k-way merging pattern if and only if:

1.   there are at most $(k + 1)$ distinct tape labels on the edges at a node;

2.   the step numbers on any root-leaf path are in descending order;

3a. for a read-forward tree—if the edges (r, s) and (t, u) have the same tape label, then $r < t$ implies $s < u$;

3b. for a read-backward tree—if the edges (r, s) and (t, u) have the same tape label, then it is not true that $r < t < s < u$.

The read-backward-merging step changes ascending strings to descending strings, and vice versa. All the edges at the same level of the read-backward tree represent either ascending strings or represent descending strings; and they alternate from level to level.

One measure of the effectiveness of a tape-merging program is the number of times the original strings are passed through the computer in the complete sort. This is the sum of the root-to-leaf distances of the merging tree. There are a number of unbalanced merging methods, which correspond to systematic methods of building merging trees in layers, or phases. In the former, the tree is grown at all endpoints with the same tape label in one phase. In a phase, strings are merged from several tapes repeatedly until one becomes empty. A *cycle* includes a number of such phases of growing the tree (or merging the tapes) before the strategy is repeated.

Associated with each merging method is the largest tree that can be grown by n cycles. The distribution of strings on the tapes after each cycle is known as a *perfect distribution* for that method. Two generating functions can be constructed that describe the distribution of the end points at each level for these *perfect* trees. The first enumerates the endpoints, or strings, in perfect distribution; the second, the weights of these perfect trees. It is possible to approximate these series and to obtain a close estimate of the power of each sorting method. The simplest unbalanced merging method is called *polyphase merging*.

Polyphase merging. The discipline for polyphase merging is to maintain a k-way merge using $(k + 1)$ tapes. The strings are first distributed on k tapes, leaving the $(k + 1)^{st}$ empty. Strings are then merged from the k tapes until one is empty. This phase leaves k tapes containing strings and one tape empty. This k-way merging is repeated until k tapes contain one string each. The final phase merges these k strings to form the final sorted string. The polyphase merging is divided into cycles having one phase per cycle.

Examples of both a table and a tree for polyphase merging of 31 strings using four tapes are given in Fig. 5.23. In each pair (i, j) of the table, i is the number of strings and j is the length of each string. The number of string transfers is 31 for the initial distribution, plus the number of strings

Phase number					Total strings	Number of strings written	Length of strings written
0	0,1 +7	7,1 −7	11,1 −7	13,1 −7	31		
1	7,3 −4	0,1 +4	4,1 −4	6,1 −4	17	7	3
2	3,3 −2	4,5 −2	0,1 +2	2,1 −2	9	4	5
3	1,3 −1	2,5 −1	2,9 −1	0,1 +1	5	2	9
4	0,5 +1	1,5 −1	1,9 −1	1,17 −1	3	1	17
	1,31	0,5	0,9	0,17	1	1	31

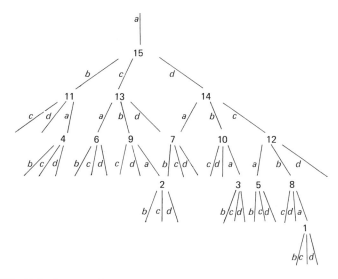

Fig. 5.23 Polyphase sorting

written at each phase times their lengths (given at the side of the table). The total number of string transfers is 138. In the following program, the total tapes are represented as a list of tapes, and a single tape is represented as a list of strings. The list of tapes is arranged in order of decreasing number of strings. A phase repeatedly merges strings using *mmerge* until the first tape becomes empty.

> **def** *phase x* = *phase*1 () *x*
>> **where rec** *phase*1 *y x* =
>> **if** *null*(*h x*)
>> **then** *y*, *t x*
>> **else** *phase*1(*postfix*(*mmerge*(*map h x*))*y*)(*map t x*)

The function *phase* produces a pair whose first is the result of merging and whose second is a list of the remaining nonempty tapes. The complete polyphase sort repeats the function *phase* until one string remains. It prefixes the result of a phase to the remaining tapes between phases.

> **def rec** *polyphase x* =
>> **let** *y, z* = *phase x*
>> **if** *all null z*
>> **then** *y*
>> **else** *polyphase*(*postfix y z*)

The number of strings on the tapes between phases for the example in Fig. 5.23 are given below:

b	*c*	*d*	*a*	*b*	*c*	*d*	*a*
7	11	13					
	4	6	7				
		2	3	4			
			1	1	2		
				1	1	1	
					0	0	1

In general the sequence of numbers for which perfect distributions exist is the generalized Fibonacci numbers, having the following recurrence relation:

$$F(n + k) = F(n + k - 1) + F(n + k - 2) + \ldots + F(n)$$

in which each term is the sum of the preceding k terms. The shapes of the first few perfect distributions for $k = 2$ are shown in Fig. 5.24. The order of the step numbers is reversed to show how the trees are grown.

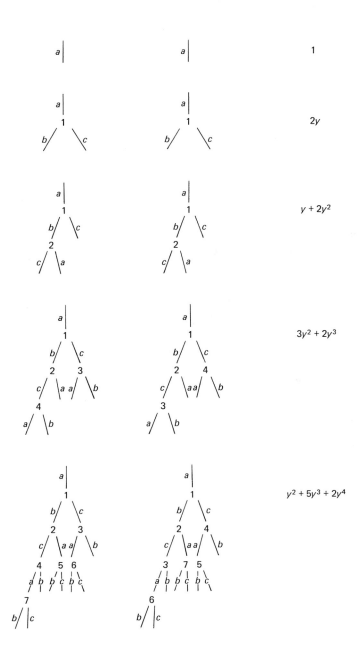

Fig. 5.24 Polyphase trees

The read-forward and read-backward trees are the same shape. The only difference is in the step numbering. The level numbers of the end points of each tree can be characterized by a two-variable generating function P_2 (x, y), in which the coefficient of $x^n y^r$ is the number of end points on the rth level of an n-phase, 2-way polyphase sort tree. The first few terms are:

$$P_2(x, y) = 1 + 2yx + (y + 2y^2)x^2 + (3y^2 + 2y^3)x^3 +$$
$$(y^2 + 5y^3 + 2y^4)x^4 + (4y^3 + 7y^4 + 2y^5)x^5 + \ldots.$$

The nth tree, (F_n) has an F_{n-1} tree as its left sub-tree and an F_{n-2} tree as its right sub-tree, and F_1 and F_2 are the first two trees in Fig. 5.24. The generating function satisfies the relation

$$P_2(x, y) = yxP_2(x, y) + yx^2P_2(x, y) + 1 + xy$$
$$P_2(x, y) = (1 + xy)/(1 - y(x + x^2))$$

Two generating functions can be obtained from $P_2(x, y)$. If y is set equal to 1, then the coefficient of x^n is the number of endpoints of the tree. If $P_2(x, y)$ is differentiated with respect to y then the number of endpoints on each level is multiplied by the level number. These are then accumulated by setting $y = 1$. The generating function for the weights of the trees is $P_2^y(x, 1)$. Thus

$$P_2(x, 1) = 1 + 2x + 3x^2 + 5x^3 + 8x^4 + 13x^5 + \ldots$$
$$= (1 + x)/(1 - x - x^2)$$
$$P_2^y(x, 1) = 1 + 2x + 5x^2 + 12x^3 + 25x^4 + 50x^5 + \ldots$$
$$= (2x + x^2)/(1 - x - x^2)^2$$

The generating function $P_2(x, y)$, and more generally $P_k(x, y)$ for a k-way polyphase merge, can be divided into component generating functions as follows. Suppose the nonempty tapes are numbered according to the number of strings on them which have perfect distributions. The tape containing the most strings is numbered 1; the tape with the next highest number of strings is numbered 2; etc. In other words suppose the list of tapes which appear as an argument to *phase* are numbered from the end. The association of tape label and number will change from one phase to the next. Suppose there is a generating function for each tape, $T_i(x, y)$ for tape number i, in which the coefficient of $x^n y^r$ is the number of endpoints for tape i on the rth level of the tree for an n-phase, k-way, polyphase merge. Thus for $k = 2$:

$$T_1 = 1 + yx + (y + y^2)x^2 + (2y^2 + y^3)x^3 + (3y^2 + 4y^3 + y^4)x^4 + \ldots$$
$$T_2 = yx + y^2x^2 + (y^2 + y^3)x^3 + (2y^2 + y^3)x^4 + \ldots$$

Consider next how a polyphase tree is grown. At each phase the tree is grown at endpoints that correspond to the tape having the greatest number of endpoints. At the end of any phase, the tape holding the greatest number of strings will have accepted contributions of strings from those tapes which held both the greatest number of strings and second greatest number of strings before the phase began. In general any tape T_i will have had a contribution both from the previous T_1 and from T_{i-1}. T_{i-1} remains undisturbed in the tree; but the new endpoints which came from T_1 are moved down one level. For $k = 2$ the generating functions can be defined as:

$$T_1 = xyT_1 + xT_2 + 1$$

and

$$T_2 = xyT_1$$

These can be solved to give

$$T_1 = 1/(1 - y(x + x^2))$$

and

$$T_2 = xy/(1 - y(x + x^2))$$

yielding the same result as before for $P_2(x, y) = T_1 + T_2$. The generating functions for a k-way polyphase merge are given by:

$$T_1 = xyT_1 + xT_2 + 1$$
$$T_2 = xyT_1 + xT_3$$
$$T_3 = xyT_1 + xT_4$$
$$\vdots$$
$$T_{k-1} = xyT_1 + xT_k$$
$$T_k = xyT_1$$

These equations can be solved to give

$$P_k(x, y) = T_1 + T_2 + \cdots + T_k = (1 + y((k - 1)x + (k - 2)x^2 + \cdots + x^{k-1}))/(1 - y(x + x^2 + \cdots + x^k))$$

If this is differentiated with respect to y, and y is then set equal to 1, the result is the generating function for the weights of the perfect trees:

$$W_k(x) =$$
$$(kx + (k - 1)x^2 + (k - 2)x^3 + \cdots + x^k)/(1 - x - x^2 - \cdots - x^k)^2$$

If $P_k(x, 1) = \Sigma p_{k,n}x^n$ and $W_k(x) = \Sigma w_{k,n}x^n$, then the number of string

k	a_k	U_k	V_k	Power
2	1.6180	1.5037	0.9920	1.9445
3	1.8393	1.0148	0.9645	2.6788
4	1.9276	0.8630	0.9206	3.1860
5	1.9659	0.7958	0.8635	3.5136
6	1.9836	0.7618	0.7965	3.7160
7	1.9920	0.7436	0.7229	3.8372
8	1.9960	0.7337	0.6459	3.9081
9	1.9980	0.7281	0.5682	3.9487
10	1.9990	0.7251	0.4920	3.9717
15	1.9997	0.7215	0.1655	3.9987
20	1.9999	0.7214	−0.0715	3.9999

Fig. 5.25   The effectiveness of polyphase sorting

transfers needed to sort $p_{k,n}$ strings is $p_{k,n} + w_{k,n}$. $P_k(x, 1)$ can be closely approximated by an expression of the form $a_k/(1 - \alpha_k)$; and $W_k(x)$ can be approximated by an expression of the form $b_k/(1 - \alpha_k x) + c_k/(1 - \alpha_k x)^2$, where α is the reciprocal of the smallest positive root of $1 - x - x^2 - \ldots - x^k = 0$.

The calculation of α_k, a_k, b_k, and c_k leads to an expression of the number of string transfers in the form $U_k S \ln S + V_k$ where S is the number of strings to be sorted. The first few values of U_k and V_k are given in Fig. 5.25. The effective power of the merge is $\exp(1/U_k)$, and approaches 4 asymptotically. This is equivalent to a balanced merged with 8 tapes. If string transfers are the sole criterion, the polyphase merge is superior to a balanced merge only when $k \leq 8$.

Cascade merging.   The discipline followed in a *cascade* merging method is to distribute the strings on all but one of the $(k + 1)$ tapes available for merging. Strings are then merged repeatedly from k tapes until one tape is empty. Then a $(k - 1)$-way merge takes place from the $(k - 1)$ unemptied tapes to the one just emptied. The process repeats, performing $(k - 2)$-way, $(k - 3)$-way merges, etc., and ends with a 2-way merge, followed by a 1-way merge (or copy). Once a tape is inscribed during a phase it plays no part in the rest of the cycle. The whole cycle of k-, $(k - 1)$-, $(k - 2)$-, $\ldots$, 1-way merging moves all the strings from one set of k tapes to another. This cycle is then repeated on each new set of k tapes, until there are k tapes, each containing one string, which are then merged to form the final sorted string. Examples of both a table and a tree for cascade merging of 31 strings using 4 tapes is given in Fig. 5.26. All steps within any particular phase are given the same step number.

The levels of the tree correspond to cycles. In each cycle all strings are moved from one set of k tapes to another set, which omits the tape which previously held the greatest number of strings. The total number of string transfers is equal to the number of levels of the tree plus one for

	a	b	c	d
	14	11	6	0
(1)	−6	−6	−6	+6
(2)	−5	−5	+5	0
(3)	−3	+3	0	0
	0	3	5	6
(4)	+3	−3	−3	−3
(5)	0	+2	−2	−2
(6)	0	0	+1	−1
	3	2	1	0
(7)	−1	−1	−1	+1
(8)	−1	−1	+1	0
(9)	−1	+1	0	0
	0	1	1	1
(10)	1	−1	−1	−1
	1	0	0	0

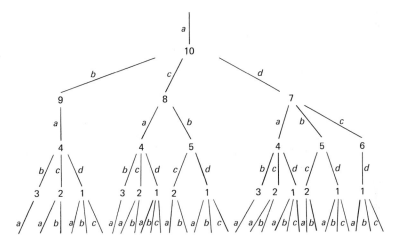

Fig. 5.26 A cascade merge table and tree

the initial distribution, all multiplied by the number of strings. In this example, the number of string transfers is $5 \times 31 = 155$. The cascade-merging discipline can be expressed as follows. The input is, as in the polyphase case, a list of tapes in ascending order of the number of strings they contain.

> **def rec** *cascadecycle x* =
>
>   **let** *y, z* = *phase x*
>
>   **if** *null z*
>
>   **then** ()
>
>   **else** *y* :(*cascadecycle z*)

Each cascade cycle leaves the tapes in descending order of the number of strings they contain, thus the list must be reversed between cycles.

> **def rec** *cascade x* =
>> **let** *z* = *cascadecycle x*
>> **if** *all null*(*t z*)
>> **then** *h z*
>> **else** *cascade*(*reverse z*)

The last phase in a cascade cycle is merely a copy operation and can be omitted, producing a strategy known as a *modified cascade merge.* The modified cascade cycle can be written as:

> **def rec** *modcascadecycle x* =
>> **let** *y, z* = *phase x*
>> **if** *null*(*t z*)
>> **then** *y*:*z*
>> **else** *y*:(*modcascadecycle z*)

If the strings shown in Fig. 5.26 are merged using the modified cascade method; then the steps (3), (6), and (9) of the table in Fig. 5.26 must be omitted, giving the table and tree in Fig. 5.27. The generating function for the cascade merge is $T_1 + T_2 + \ldots + T_k$ where the T_i are related as follows:

$$T_1 = xy(T_1 + T_2 + \ldots + T_k) + 1$$
$$T_2 = xy(T_1 + T_2 + \ldots T_{k-1})$$

$$\vdots$$

$$T_r = xy(T_1 + T_2 + \ldots + T_{k+1-r})$$

$$\vdots$$

$$T_{k-1} = xy(T_1 + T_2)$$
$$T_k = xyT_1$$

and for modified cascade the only change is to T_1:

$$T_1 = xy(T_1 + T_2 + \ldots + T_{k-1}) + T_k + 1$$

The cascade merge equations can be solved, using Cramer's rule, to give

$$T_1 = h_{k-2}(xy)/h_k(xy)$$

	a	*b*	*c*	*d*
	14	11	6	0
(1)	−6	−6	−6	+6
(2)	−5	−5	+5	0
	3	0	5	6
(3)	−3	+3	−3	−3
(4)	+2	0	−2	−2
	2	3	0	1
(5)	−1	−1	+1	−1
(6)	−1	−1	0	+1
	0	1	1	1
(7)	+1	−1	−1	−1
	1	0	0	0

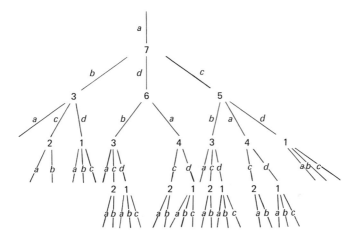

Fig. 5.27 A table and tree for the modified cascade merge

where

$$h_k = \det(\delta_{ij} - a_{ij})$$

and

$$a_{ij} = \begin{cases} 0 \text{ if } i + j > k + 1 \\ 1 \text{ if } i + j \le k + 1 \end{cases}$$

These equations have the recurrence relation.

$$h_k(x) = (-1)^{n-1} x h_{k-1}(-x) + h_{k-2}(x)$$
$$h_0 = 1, h_1 = 1 - x.$$

If x is now set to $2 cos\, y$, the solution can be found to be [5–32]:

$$h_k = (-1)^{n(n+1)/2}\sin((k + 1/2)y(-1)^k)/\sin(y/2)$$

The roots of $h_k = 0$ are given by $(k + 1/2)y = m\,\pi$ and are

$$(-1)^k 2\, cos(2m\,\pi/(2k + 1))\, for\, m = 1, 2, 3, \ldots, k$$

The smallest positive root is $2cos(k\,\pi/(2k + 1))$, which is approximately equal to $\pi/(2k + 1)$. The reciprocal of this is the effective power of the k-way cascade merge.

The sequence of values for which perfect cascade distributions exist is closely approximated by an expression of the form $C(k, n) = a_k((2k + 1)/\pi)^n$. The number of string transfers needed to sort $N(k, n)$ strings is $n \times N(k, n)$, or $(n\, \ln\, N - \ln\, a_k)/\ln((2k + 1)/\pi)$. The number of string transfers needed to sort S items by a cascade merge is closely approximated by $U_k S\, \ln\, S + V_k$, where U_k and V_k are given in Fig. 5.28.

	Effective power cascade	Cascade	
k	α_k	U_k	V_k
2	1.6180	2.0781	0.6723
3	2.2470	1.2352	0.7540
4	2.8794	0.9456	0.7957
5	3.5133	0.7958	0.8213
6	4.1481	0.7029	0.8388
7	4.7834	0.6389	0.8515
8	5.4190	0.5917	0.8613
9	6.0548	0.5553	0.8690
10	6.6907	0.5261	0.8754
15	9.8718	0.4367	0.8954
20	13.0539	0.3892	0.9064

Fig. 5.28   Efficiency of cascade merging

Compromise merging techniques.   The cascade, modified-cascade, and polyphase methods are special cases of a family of *compromise* merging methods, so called because they are a compromise between the cascade and polyphase methods. The cascade technique uses phases of k (,)- $(k-1)$-, $(k-2)$-, $\ldots$, 1)-way merges, the polyphase method uses k-way merging throughout. A compromise merging method first has a phase of k-way merging, then a phase of $(k-1)$-way merging, as in the cascade technique. The compromise method, however, stops its cycle after a $(k - r + 1)$-way merge. At this stage the cycle is repeated. A compromise merge is governed by two parameters (k, r). A $(k, 1)$ compromise merge is polyphase, (k, k) is cascade, and

$(k, k - 1)$ is modified cascade. The programs for compromise merging are:

def rec *compromisecycle r x =*

       if $r = 0$

       then (*reverse x*)

       else let $y, z = $ *phase x*

                         y:(*compromisecycle*$(r - 1)z$)

def rec *compromise r x =*

       let $z = $ *compromisecycle r x*

       if *all null* ($t z$)

       then $h z$

       else *compromise r* (*reverse z*)

Examples of both a table and a tree for a (5, 2) compromise merge of 33 strings are shown in Fig. 5.29. In this case only 83 string transfers are required. This is compared with 88 for a (5, 1) polyphase sort of 33 items.

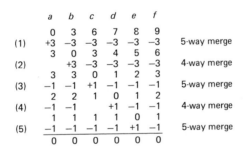

	a	b	c	d	e	f	
	0	3	6	7	8	9	
(1)	+3	−3	−3	−3	−3	−3	5-way merge
	3	0	3	4	5	6	
(2)		+3	−3	−3	−3	−3	4-way merge
	3	3	0	1	2	3	
(3)	−1	−1	+1	−1	−1	−1	5-way merge
	2	2	1	0	1	2	
(4)	−1	−1		+1	−1	−1	4-way merge
	1	1	1	1	0	1	
(5)	−1	−1	−1	−1	+1	−1	5-way merge
	0	0	0	0	0	0	

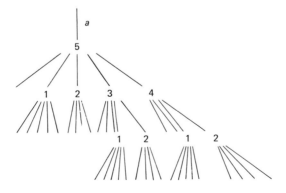

Fig. 5.29 An example of a (5,2) compromise merge

The relative efficiencies of the compromise methods can be obtained by a technique similar to that used for the polyphase and cascade methods. The generating function for the (k, r) compromise merge is $T_1 + T_2 + \ldots + T_k$ where

$$T_1 = xy(T_1 + T_2 + \ldots + T_r) + T_{r+1} + 1$$
$$T_2 = xy(T_1 + T_2 + \ldots + T_r) + xT_{r+2}$$

$$\vdots$$

$$T_{k-r} = xy(T_1 + T_2 + \ldots + T_r) + xT_k$$
$$T_{k-r+1} = xy(T_1 + T_2 + \ldots + T_r)$$
$$T_{k-r+2} = xy(T_1 + T_2 + \ldots + T_{r-1})$$

$$\vdots$$

$$T_{k-1} = xy(T_1 + T_2)$$
$$T_k = xyT_1$$

The number of string transfers can be put into the form

$$U_{kr}S \ln S + V_{kr}S$$

and the effective power is $\exp(1/U_{kr})$. These three constants, U_{kr}, V_{kr}, and the effective power, are tabulated in Figs. 5.30 to 5.32. As k tends to infinity, the effective power of a (k, r) compromise merge tends to $(r + 1)^{((r+1)/r)}$. From the tables it can be seen that the best values of r for each k are:

k	2	3	4	5	6	7	8	9	10
r	1	1	2	2	3	4	6	7	7

r	1	2	3	4	5	6	7	8	9	Cascade
k 2	1.5037	2.0781								
3	1.0148	1.0232	1.2352							
4	0.8630	0.8521	0.8963	0.9456						
5	0.7958	0.7475	0.7563	0.7728	0.7958					
6	0.7618	0.6911	0.6747	0.6815	0.6905	0.7029				
7	0.7436	0.6583	0.6283	0.6255	0.6258	0.6315	0.6389			
8	0.7337	0.6393	0.5992	0.5855	0.5834	0.5831	0.5870	0.5917		
9	0.7281	0.6269	0.5806	0.5591	0.5532	0.5494	0.5493	0.5521	0.5553	
10	0.7251	0.6196	0.5676	0.5409	0.5296	0.5248	0.5218	0.5218	0.5239	0.5261
15	0.7215	0.6080	0.5448	0.5051	0.4797	0.4627	0.4529	0.4463	0.4411	0.4367
20	0.7214	0.6969	0.5415	0.4985	0.4683	0.4465	0.4299	0.4184	0.4104	0.3892

Fig. 5.30   Values of U for compromise merging

r	1	2	3	4	5	6	7	8	9	Cascade
k 2	0.9920	0.6723								
3	0.9645	0.8199	0.7540							
4	0.9206	0.8826	0.7994	0.7957						
5	0.8635	0.8879	0.8318	0.8080	0.8213					
6	0.7965	0.8681	0.8585	0.8304	0.8213	0.8388				
7	0.7229	0.8491	0.8712	0.8400	0.8374	0.8341	0.8515			
8	0.6459	0.8067	0.8612	0.8503	0.8448	0.8457	0.8452	0.8615		
9	0.5682	0.7696	0.8357	0.8623	0.8461	0.8525	0.8535	0.8545	0.8690	
10	0.4921	0.7198	0.8188	0.8588	0.8481	0.8535	0.8598	0.8606	0.8624	0.8754
15	0.1655	0.4822	0.6529	0.7562	0.8092	0.8488	0.8544	0.8584	0.8721	0.8954
20	−0.0715	0.2821	0.4860	0.6182	0.7084	0.7710	0.8161	0.8439	0.8543	0.9064

Fig. 5.31 Values of V for compromise merging

r	1	2	3	4	5	6	7	8	9	Cascade
k 2	1.9445	1.6180								
3	2.6788	2.4773	2.2470							
4	3.1860	3.2333	3.0513	2.8794						
5	3.5136	3.8103	3.7521	3.6470	3.5133					
6	3.7160	4.2501	4.4022	4.3379	4.2553	4.1481				
7	3.8372	4.5675	4.9122	4.9468	4.9426	4.8715	4.7834			
8	3.9081	4.7790	5.3065	5.5171	5.5522	5.5558	5.4928	5.4190		
9	3.9487	4.9296	5.5972	5.9808	6.0957	6.1724	6.1743	6.1177	6.0548	
10	3.9717	5.0225	5.8227	6.3526	6.6039	6.7220	6.7978	6.7962	6.7450	6.6907
15	3.9987	5.1798	6.2684	7.2423	8.0410	8.6818	9.0991	9.3990	9.6486	9.8718
20	3.9999	5.1947	6.3391	7.4323	8.4592	9.3915	10.2392	10.9165	11.4373	13.0539

Fig. 5.32 The effective powers of compromise merging

REFERENCES

Most of the sorting methods have been reexpressed here in a functional notation can be found in the extensive literature on sorting. Surveys and bibliographies of sorting methods can be found in references 5-1, 5-3, 5-9, 5-10, 5-15, 5-17, 5-22, 5-24, 5-25, 5-28, 5-34, and 5-35. The methods treated in this chapter can be found in the references as follows:

Sorting networks—5-2, 5-4, 5-13

Shell sort—5-30, 5-36

Binary search trees—5-18, 5-26, 5-39

Quicksort—5-20, 5-21

Radix exchange—5-19

Heapsort—5-11, 5-12, 5-38

Tape sorting—5-5, 5-6, 5-8, 5-23, 5-27, 5-31, 5-33, 5-37

5-1. Ashenhurst, R. L., "Sorting and arranging," (in) *Theory of Switching, Report No. BL-7,* Harvard Comput. Lab., May 1954, pp. I.1–I.76.

5-2. Batcher, K. E., "Sorting networks and their applications," *Proc. AFIPS, 1968 SJCC,* Vol. 32, Montvale, N.J.: AFIPS Press, pp. 307–314.

5-3. Bell, D. A., "The principles of sorting," *Computer Journal,* Vol. 1, 1958, pp. 71–77.

5-4. Bose, R. C., and R. J. Nelson, "A sorting problem," *JACM,* Vol. 9, No. 2, 1962, pp. 282–296.

5-5. Burge, W. H., "Sorting, trees, and measures of order," *Information and Control,* Vol. 1, 1958, pp. 181–197.

5-6. Burge, W. H., "An analysis of the compromise merge sorting techniques," *Proc. IFIP Cong. 1971,* Amsterdam: North Holland, 1972, pp. 454–459.

5-7. Burge, W. H., "An analysis of a tree sorting method and some properties of a set of trees," *Proc. 1st USA-Japan Computer Conference,* 1972, pp. 372–379.

5-8. Carter, W. C., "Mathematical analysis of merge-sorting techniques," *Proc. IFIP Cong. 1962,* Amsterdam: North Holland, 1963, pp. 62–66.

5-9. Davies, D. W., "Sorting of data on an electronic computer," *Proc. Inst. Elec. Eng. 103B,* 1956, Supplement 1, pp. 87–93.

5-10. Flores, I., *Computer Sorting,* Englewood Cliffs, N.J.: Prentice-Hall, 1969.

5-11. Floyd, R. W., "Treesort: Algorithm 113," *CACM,* Vol. 5, No. 8, 1962, p. 434.

5-12. Floyd, R. W., "Treesort: Algorithm 245," *CACM,* Vol. 7, No. 12, 1964, p. 701.

5-13. Floyd, R. W., and D. E. Knuth, "The Bose-Nelson sorting problem," (in) *A Survey of Combinatorial Theory,* J. N. Srivastava, et al. (eds.), Amsterdam: North Holland, 1973, pp. 163–172.

5-14. Ford, L. R., and S. M. Johnson, "A tournament problem," *Amer. Math. Monthly,* Vol. 66, 1959, pp. 387–389.

5-15. Friend, E. H., "Sorting on electronic computer systems," *JACM,* Vol. 3, 1956, pp. 134–168.

5-16. Gale, D., and R. Karp, "A phenomenon in the theory of sorting," *IEEE Conf. Record of the 11th Annual Symposium on Switching and Automata Theory,* 1970, pp. 51–59.

5-17. Gotlieb, Calvin C., "Sorting on computers," *CACM,* Vol. 6, No. 5, 1963, pp. 194–201.

5-18. Hibbard, T. N., "Some combinatorial properties of certain trees with applications to sorting and searching," *JACM,* Vol. 9, 1962, pp. 13–28.

5-19. Hildebrandt, P., and H. Isbitz, "Radix exchange—an internal sorting method for digital computers," *JACM,* Vol. 6, 1959, pp. 156–163.

5-20. Hoare, C. A. R., "Partition, Quicksort and Find (Algorithms 63, 64 and 65)," *CACM,* Vol. 4, No. 7, 1961, p. 321, Vol. 6, No. 8, 1963, p. 446.

5-21. Hoare, C. A. R., "Quicksort," *Computer Journal,* Vol. 5, 1962, pp. 10–15.

5-22. Hosken, J. C., "Evaluation of sorting methods," *Proc. Eastern JCC,* Vol. 8, 1955, pp. 39–55.

5-23. Knuth, D. E., "Letters to the editor regarding the ACM glossary of sorting and merging terms," *CACM,* Vol. 6, No. 10, 1963, pp. 585–587.

5-24. Knuth, D. E., *Sorting and Searching, The Art of Computer Programming,* Vol. 3, Reading, Mass.: Addison-Wesley, 1973.

5-25. Lorin, H. A., "A guided bibliography to sorting," *IBM Systems J.,* 1971, pp. 244–254.

5-26. Lynch, W. C., "More combinatorial properties of certain trees," *Computer Journal,* Vol. 7, 1964, pp. 299–302.

5-27. Lynch, H. A., "The t-fibonacci numbers and polyphase sorting," *Fib. Quart.,* Vol. 8, 1970, pp. 6–22.

5-28. Martin, W. A., "Sorting," *Computing Surveys,* Vol. 3, 1971, pp. 147–174.

5-29. Peterson, W. W., "Bouricius' theorem on sorting devices," *IBM Research Report* IR-00051, 1956.

5-30. Pratt, V., "Shell sorting and sorting networks," Ph.D. Thesis, Stanford University, 1971.

5-31. Radke, C. E., "Merge-sort analysis by matrix techniques," *IBM Systems Journal,* Vol. 5, 1966, pp. 226–247.

5-32. Raney, G. N., "Generalization of the Fibonacci sequence to n dimensions," *Canadian J. Math.,* Vol. 18, 1966, pp. 332–349.

5-33. Reynolds, S. W., "A generalized polyphase merge algorithm," *CACM,* Vol. 4, 1961, pp. 347–349.

5-34. Rich, R. P., *Internal Sorting,* Englewood Cliffs, N.J.: Prentice-Hall, 1973.

5-35. Rivest, R. L., and D. E. Knuth, "Computer sorting, bibliography 26," *Computing Rev.,* Vol. 13, 1972, pp. 283–289.

5-36. Shell, D. L., "A high-speed sorting procedure," *CACM,* Vol. 2, No. 7, 1959, pp. 30–32.

5-37. Sobel, S., "Oscillating sort—a new sort merging technique," *JACM,* Vol. 9, 1963, pp. 372–374.

5-38. Williams, J. W. J., "Heapsort: Algorithm 232," *CACM,* Vol. 7, 1964, pp. 347–348.

5-39. Windley, P. F., "Trees, forests, and rearranging," *Computer Journal,* Vol. 3, 1960, pp. 84–88.

ABOUT THE AUTHOR

Mr. Burge is a Research Staff member at the I.B.M. T. J. Watson Research Center, Yorktown Heights, New York.

He was first initiated into programming on the TREAC, a copy of the EDSAC computer at the Royal Radar Establishment, Great Malvern, England, during the summer of 1954. After graduating from Cambridge University with a degree in Mathematics in 1955, he became a research assistant at U.C.L.A. working on numerical analysis programs for the SWAC computer.

On returning to England in 1956, he joined E.M.I. Electronics Ltd. where he worked on the design of the EMIDEC 2400 computer and its programming systems. During that time he became interested in the use and implementation of LISP and ALGOL 60 and wrote an ALGOL 60 compiler for the EMIDEC 2400. In 1963 he joined the UNIVAC Division of the Sperry Rand Corporation in New York City as Manager of Systems Programming Research. There he was head of a small group that was attempting to formulate notions of the semantics of programming languages. Since 1965, he has been at the I.B.M. Research Center studying methods of simplifying programming by using high-level programming languages and methods for the combinatorial analysis of algorithms. He is the author of many papers on sorting methods and programming languages.

INDEX TO PROGRAMS

INDEX